The Rationalists: Between Tradition and Innovation

The New Synthese Historical Library
Texts and Studies in the History of Philosophy

VOLUME 65

For further volumes:
http://www.springer.com/series/6608

Carlos Fraenkel · Dario Perinetti · Justin E.H. Smith
Editors

The Rationalists: Between Tradition and Innovation

Springer

Editors
Prof. Carlos Fraenkel
Department of Philosophy
McGill University
Leacock Building
855 Sherbrooke Street
West Montréal, Québec H3A 2T7
Canada
carlos.fraenkel@mcgill.ca

Prof. Dario Perinetti
Département de Philosophie
Université du Québec à Montréal
Case postale 8888, succursale Centre-ville
Montréal, Québec H3C 3P8
Canada
perinetti.dario@uqam.ca

Prof. Justin E.H. Smith
Department of Philosophy
Concordia University
1455 Blvd. de Maisonneuve
West Montréal, Québec H3G 1M8
Canada
justismi@alcor.concordia.ca

ISBN 978-90-481-9384-4 e-ISBN 978-90-481-9385-1
DOI 10.1007/978-90-481-9385-1
Springer Dordrecht Heidelberg London New York

Library of Congress Control Number: 2010936524

Printed on acid-free paper

Springer is part of Springer Science+Business Media (www.springer.com)

Contents

Contributors

Roger Ariew Department of Philosophy, University of South Florida, Tampa, FL, USA, rariew@cas.usf.edu

François Duchesneau Université de Montréal, Montréal, QC, Canada, francois.duchesneau@montreal.ca

Carlos Fraenkel Department of Philosophy, McGill University, Leacock Building, 855 Sherbrooke Street, West Montréal, Québec H3A 2T7, Canada, carlos.fraenkel@mcgill.ca

Alison Laywine McGill University, Montreal, QC, Canada, a.laywine@mcgill.ca

Brandon C. Look University of Kentucky, Lexington, KY, USA, look@uky.edu

Syliane Malinowski-Charles Bishop's University, Sherbrooke, QC, Canada, scharles@ubishops.ca

Yitzhak Y. Melamed Johns Hopkins University, Baltimore, MD, USA, ymelame1@jhu.edu

Ohad Nachtomy Bar-Ilan University, Ramat Gan, Israel; Fordham University, New York, NY, USA, ohnadnachtomy@mac.com

Steven Nadler University of Wisconsin-Madison, Madison, WI, USA, smnadler@wisc.edu

Dario Perinetti Département de Philosophie, Université du Québec à Montréal, Case postale 8888, succursale Centre-ville, Montréal, Québec H3C 3P8, Canada, perinetti.dario@uqam.ca

Lisa Shapiro Simon Fraser University, Burnaby, BC, Canada, lshapiro@sfu.ca

Hasana Sharp McGill University, Montreal, QC, Canada, hasana.sharp@mcgill.ca

Justin E.H. Smith Department of Philosophy, Concordia University, 1455 Blvd. de Maisonneuve, West Montréal, Québec H3G 1M8, Canada, justismi@alcor.concordia.ca

Chapter 1
Introduction

Carlos Fraenkel, Dario Perinetti, and Justin E.H. Smith

According to Francis Bacon, "the Rationalists are like spiders who weave webs out of their own bodies." The "Empiricists," by contrast, "are like ants who only collect things and make use of them."[1] This amusing characterization notwithstanding, it is no longer a secret that the distinction between Rationalists and Empiricists in the early modern period, both of which movements came long after Bacon's death, is in many ways an anachronistic projection back onto two loose fraternities of thinkers who certainly did not define themselves in these terms, and generally did not perceive themselves as having common cause with the other philosophers with whom they would be posthumously grouped. The habit of carving up early modern philosophers in this way appears to have emerged only in Kant's time and reflects most notably the nineteenth century fondness for periodizations and for the retroactive placing of historical figures within schools.[2]

Today the distinction between Rationalism and Empiricism is often acknowledged as pedagogically useful, but as failing to capture the complexities of the philosophical concerns and orientations of the thinkers it is meant to characterize. Yet no one is quite ready to dispense with it, and scholarly works continue to be published, including this one, that feature the label in their titles.[3]

The attempt to come up with a set of core doctrines, shared by all philosophers traditionally grouped together as the seventeenth-century Rationalists, is a

C. Fraenkel (✉)
Department of Philosophy, McGill University, Leacock Building, 855 Sherbrooke Street, West Montréal, Québec H3A 2T7, Canada
e-mail: carlos.fraenkel@mcgill.ca

[1]Francis Bacon, *Novum Organum*, Aphorism 1.95.

[2]Dominik Perler, following John Cottingham, identifies the division between Empiricists and Rationalists as dominating philosophical discussions throughout the eighteenth century. This is certainly true with respect to the content of many discussions, but does not mean that the participants in these debates organized themselves under these descriptions (Perler 1998, 252f). For a critique of this way of construing the history of eighteenth century philosophy and an account of its origins in post-Kantian historiography of philosophy, see Haakonssen (2006), and also Schneewind (2004).

[3]See in particular Cottingham (1998), Gennaro and Huenemann (2002), Phemister (2006), and Huenemann (2008).

C. Fraenkel et al. (eds.), *The Rationalists: Between Tradition and Innovation*, The New Synthese Historical Library 65, DOI 10.1007/978-90-481-9385-1_1,

considerable challenge. Not even the belief that every thing or event in the world has a reason for its being, that the world is, in short, a rational place, can be confidently attributed to all of them. There is, for instance, substantial evidence that for Descartes all things, including necessary truths in mathematics and logic, are contingent, i.e., dependent on God's will and could be otherwise if God had willed so.[4] This led Leon Roth to compare Descartes's position to the occasionalism of Asharite *Kalâm*, a medieval Arabic school of theology, and to contrast it with what he took to be the genuine rationalism of Maimonides and Spinoza.[5]

Or take the related belief that all real knowledge is arrived at by intellectual intuition or the use of reason and deduction, rather than inductively by use of the senses. Descartes for one appears to accept the authority of a third source of knowledge. In *Principia Philosophiae* 1.76, for instance, he contends that revelation is, in fact, our most certain path to knowledge and that in case of a contradiction between reason and revelation, the former must submit to the latter:

> But above all we must imprint in our memory as the highest rule that the things revealed by God to us [*nobis a Deo revelata*] must be believed as more certain than everything [*ut omnium certissima esse credendam*]. And however strongly the light of reason appears to suggest to us something else, even if it is most clear and evident, our faith should be put in the sole divine authority, rather than in our own judgment.[6]

Despite these caveats, however, there are fairly good textual grounds for attributing the doctrine that the world is a rational place to many of the major thinkers we classify as "Rationalists", though it is Leibniz, with his principle of sufficient reason, whose philosophy is most solidly rationalistic in this sense. As for the epistemological stance traditionally associated with Rationalism, it is certainly true that, when Descartes, Spinoza, and Leibniz engage directly with questions concerning knowledge, or explicitly contrast their own views with those of empirically-minded thinkers (as, most famously, in Leibniz's 1704 response to Locke's *Essay concerning Human Understanding*), they do emphasize the importance of the faculty of reason as more fundamental than, and explanatorily prior to, anything derived from the senses. But one thing such a characterization fails to capture is the intense interest of all of the Rationalists in knowledge obtained from observation and experience, and particularly from scientific investigation of the natural world. Descartes certainly did not believe that one could deduce by a priori means the particular structure of the optic nerve. His epistemological commitments were, in fact, perfectly compatible with procuring the eyes and brains of slaughtered animals from the local butcher in order to investigate this structure.

But even if the "Rationalist" label points to something like a family resemblance between a number of seventeenth-century philosophers, we may still ask whether in grouping them under this banner we are in fact homing in on the most

---

[4]See, e.g., Resp. VI; Letters to Mersenne, 15 April 1630; 6 May 1630; 27 May 1630.

[5]See Roth (1924).

[6]AT VIII.1, 39. The view expressed in this passage is close to what Spinoza describes as "skepticism" in his *Theological-Political Treatise*, Chapter 15.

salient, or interesting, aspect of their vast contributions to philosophy and science. The Rationalists are often taken as being primarily concerned with metaphysical and epistemological questions, while short shrift is given to their interest in politics, religion, experimental science, and other fields (perhaps with the exception of scholarly attention to Spinoza's and Hobbes's political philosophies). It is worth asking whether characterizing the Rationalists *as* Rationalists does not have the effect, desired or not, of obscuring from view the entire scope of their thought. The Rationalists were rationalists, but so were many philosophers before them, and the Rationalists, like their predecessors, were many other things besides. In recent scholarship, this standard account of Rationalism has been revised for individual thinkers associated with the school, but the movement as a whole has yet to be treated in keeping with the recent turn in history-of-philosophy scholarship towards greater sensitivity for historical contexts and towards considering the full range of intellectual concerns of past thinkers in order to understand their philosophical projects.[7]

If we agree with scholars such as Jonathan Bennett, that what is of most value in the works of philosophers such as Descartes, Spinoza, and Leibniz are the arguments that we may extract from these works and, regardless of our time or place, put to use for the purpose of resolving our own philosophical problems, then the "Rationalist" label appears warranted, since the parts of these works most amenable to extraction and reconstruction concern questions of knowledge, its acquisition, and its justification.[8] But in the last few decades the English speaking world has witnessed an intense reexamination of the relationship of philosophy to its history, and of the proper methods of investigating that history, and nowhere has this rethinking been more fruitful than in the scholarly study of early modern philosophy. While until fairly recently seventeenth-century philosophers were held to be of interest to contemporary philosophy principally to the extent that their arguments could be used for solving current philosophical problems, today many philosophers studying Descartes, Spinoza, Leibniz and others have set themselves a different task: to arrive at a clearer picture of the full range of problems that interested these thinkers themselves. Certainly, rational reconstruction remains an important part of history-of-philosophy scholarship—and, we note, on the continent, and particularly in the once history-laden German tradition, it seems to be gaining ground precisely as it is retreating elsewhere—but the exclusive correctness of this methodological approach is no longer taken for granted within the discipline.[9]

The growing trend in Anglophone historiography of philosophy is to acknowledge the necessity of gaining a wider-scoped view of the work of philosophers in history, not just because such a view is truer to who these philosophers were as

---

[7] See, e.g., Garber (1992), which shows the central importance of scientific questions for Descartes's philosophy.

[8] See, e.g., Bennett (2001).

[9] For a good group of texts addressing the current methodological discussion see Zarka (2001). See also Sorell and Rogers (2005).

historically and socially embedded actors, but also, and much more importantly, because such a view is necessary for adequately understanding the core philosophical doctrines that earlier historians of philosophy were generally content to read simply as timeless exemplars of good argumentation. To cite one telling example, it is well known that over the course of the 1690s, Leibniz's mature metaphysics of substance began to take shape, and that in this metaphysics the key notions of force, both active and passive, played a central role. The notion of force can be traced back to his science of "mechanics", which starting in the 1680s Leibniz began to see as something like a bridging science between physics and metaphysics. Now, it turns out that Leibniz's notion of force took shape in a period in which he was thinking intensely not just about mechanics as an abstract theory of the motions of bodies, but in particular about the mechanics of mining: he spent the better part of the early 1680s in the silver mines of the Harz mountains, trying to figure out a way to harness the power of the wind in order to make water flow upwards, which in turn would generate all the power necessary for a major mining operation. Recent studies have argued, in steps too elaborate to describe here, that it was in the context of Leibniz's engagement with the mechanics of mining that he came to have the insights into mechanics in general that in turn brought him to his mature metaphysics of substance.[10] Now, it is not just because when Leibniz first devised his mechanics he probably thought of himself more as a mining engineer than as a metaphysician that we today should take this bit of context seriously, but also because we might discover that certain of the features of Leibniz's deepest theoretical account of the origins of motion in bodies arose out of the particular exigencies of the engineering project he had taken on.

Of course, a thorough contextualism would pay attention not just to the context of discovery, but also to the creation of traditions, institutional practices and to the particular way in which a philosopher's ideas and texts are often appropriated for purposes which that philosopher himself could not have predicted. There are as many examples of such appropriations as there are philosophers with any following, and they include, for example, Cartesian physiology in the late seventeenth century, libertine Spinozism (or the suspicion of it) in the Enlightenment, and the academic strain of Leibnizianism heralded by Christian Wolff and denounced some decades later by Kant as "dogmatism". Another example of the posthumous reception history of these three philosophers is indeed the creation of a school of "Rationalists", as well as the various projects in contemporary philosophy conceived by their authors as neo-Rationalist or as inspired by core principles of Rationalist philosophy.[11]

In addition to the myriad interests of the Rationalists and to the ways in which their thought was taken up by subsequent schools, a thorough contextualism would also attempt to shed light on the premodern roots of their thought. For the most part, the standard historiography of philosophy has taken Descartes's philosophy as the starting point of the movement, neglecting to explore its manifold relations

---

[10] See in particular Elster (1975) and Cohen and Wakefield (2008).

[11] See, e.g., Bonjour (1998).

to earlier intellectual currents.[12] Yet all of the Rationalists had significant connections to their ancient, medieval, and Renaissance predecessors.[13] One reason why these connections have been largely overlooked is that scholars often have taken Descartes's claims to have effected a radical break with the past and to have started his philosophy from scratch, at face value, rather than as the literary conceit, necessary for the narrative flow of the *Meditations*, that they are. Descartes was educated by Jesuits, and knew well the body of texts that his educators took to be canonical. Perhaps the most provocative revisionist thesis in this context has been proposed by Menn (1998a), who argued that the key to understanding Descartes's project in the *Meditations* lies in his appropriation of Augustine's method for establishing knowledge of God and the soul independently of any theory of the physical world. Descartes's lack of interest in citing his sources reveals something of the spirit of the times, and may in some ways be relevant to understanding both the method and the content of his philosophical project. But he is, of course, engaged in a dialogue with a wide range of earlier and contemporary thinkers, and the more we learn about this dialogue, the more the school of Rationalism appears to move towards the pole of tradition in the dichotomy between tradition and innovation. At least as complicated as Descartes's relationship to his Jesuit background is Spinoza's relationship to his Jewish background. Although Spinoza presents his main philosophical work, the *Ethics*, as if it were a creation out of nothing but logical inferences from definitions and first axioms, it clearly is the result of a complex critical exchange with past and contemporary intellectual traditions. Warren Zev Harvey, for instance, proposed that it is possible to read Spinoza as a Maimonidean, arguing that

> fundamental elements of Maimonides' philosophy recur as fundamental elements of [Spinoza's] philosophy. This is true [. . .] with regard to questions of psychology, epistemology, ethics, anthropology, politics, metaphysics, and true religion; that is with regard to Spinoza's philosophy as a whole, including his speculations about God and the true worship of him. (Harvey 1981, 151–72)

Although we must not make the mistake of holding seventeenth century thinkers up to our current standards of academic integrity as concerns the citing of sources, we certainly need not take them at their word. Quite the contrary, the claim of a philosopher like Descartes to be saying something never before said should be taken as an invitation to dig deeper and discover the rich network of links to his predecessors and contemporaries that, at least in his published works, he often seems to prefer to keep hidden.

Moreover, the extent to which the work of the "Rationalists" overlapped with other contemporary currents of thought, such as Cambridge Platonism, experimental philosophy à la Bayle and Oldenburg, the alchemical and Paracelsian traditions represented in the seventeenth century by figures such as Jan-Baptist van Helmont,

[12] For some significant exceptions to this, see Ariew (1998) and Des Chene (2000).

[13] See for example Stephen Menn's very useful account of the ancient, medieval, and Renaissance traditions that shaped the intellectual setting of seventeenth century philosophy (1998b).

and traditions of religious thought such as that of the Jesuits, tended to be neglected. The exact studies of the Rationalists' "minor" contemporaries, in what we would think of as nonfoundational domains of science, including chemistry, medicine, and acoustics, all help to bring into clearer focus the philosophical projects of the era's major thinkers.

Contextualism, therefore, suggests that the Rationalists were a significantly less homogeneous and distinctive group than traditionally assumed. Even the spirit of the *novatores* is not a feature consistently shared by all of them. Quite often, in fact, they do not seem interested in presenting themselves as radically new. Spinoza, for instance, in Letter 76, famously denies laying claim to originality: "I do not claim to have found the best philosophy, but I know that I understand the true one." He even contends that his concept of God as *causa immanens* agrees with a wide range of ancient sources:

> All things, I say, are in God and move in God, and this I affirm with Paul and perhaps with all ancient philosophers, though expressed in a different way, and I would even venture to say, with all the ancient Hebrews, as far as may be conjectured from certain traditions, even if these have suffered manifold corruption. (Letter 73 [Gebhardt 4.307])

Leibniz, too, is often concerned with presenting his philosophy as the true interpretation of ancient sources.

While we can say, with some qualifications, that Descartes, Spinoza, and Leibniz were Rationalists in the sense of trying to found knowledge on a few bedrock truths inhering in the rational soul prior to experience, they were a number of other things as well: scriptural exegete, empirical scientist, diplomat, and anatomist are just some of the roles we see these philosophers taking on. Much history-of-philosophy scholarship has tended to cast these sundry activities as side projects, and as inessential to understanding the philosophers themselves. In our volume we seek to draw a balanced picture by bringing the Rationalists' full range of interests, their historical antecedents, and their posthumous legacies into sharper focus, without neglecting the core commitments to the epistemological doctrine that earned them their traditional label.

The present volume has grown out of the meetings of the Montreal Interuniversity Workshop in the History of Philosophy, which has been operating since 2004, and has by now hosted a large number of papers on philosophers and philosophical currents from the pre-Socratics to the eighteenth century. The intellectual life of Montreal—a city perched, culturally if not geographically, between Europe and America—benefits from the coexistence of various scholarly traditions, and nowhere is this more visible than in the liberal approach that reigns in history-of-philosophy scholarship. The Montreal Workshop has strived to preserve as much as possible the open-mindedness informing this liberal approach. Our assumption is that a range of methodological perspectives enriches our understanding by bringing different, but not incompatible, aspects of past philosophical traditions into focus. The heterogeneity of the papers in this volume reflects our non-dogmatic stance in terms of method.

While we thus consider heterogeneity a virtue, nonetheless two overarching concerns inform, to greater and lesser extents, the contributions to the present volume. First, many of the authors are interested in critically examining the usefulness of "Rationalism" as the banner under which to group the philosophers treated. Second, the contributors intend to bring out philosophical and historical links to a wide range of intellectual contexts and concerns that significantly shaped the Rationalists' projects, but have received little attention in the way they were traditionally understood. Although, as noted, we have not imposed a uniform methodology on the approaches of the contributors to the present volume, in general the Workshop, on which the volume is based, leans strongly towards contextualism, the revision of received periodizations, and scholarship that combines philosophical argument with historical precision.

In the first section, *Continuities between the Premodern and the Modern*, the emphasis is on the extent to which Rationalist philosophy is a development out of, rather than a radical departure from, earlier traditions. In the first paper, Lisa Shapiro argues for the importance of ancient eudaimonistic ethics for understanding Descartes's ethical project. She examines in particular whether Descartes can be called a "structural eudaimonist," i.e., one who ties the human good to human nature. Although Descartes subscribes to a virtue ethics, his dualism, according to Shapiro, precludes his being a structural eudaimonist. While he offers an account of the good of the human mind as the pursuit of knowledge, and an account of the good of the human body, derived from the Stoic notion of *oikeiosis*, it is not clear how these two strands come together to form an account of the human good proper to the union of mind and body. Next, Carlos Fraenkel provides a significant example of the remarkable extent to which Spinoza—who is often presented as having laid the foundations of modernity—shared the philosophical and religious concerns of a number of his medieval predecessors in the Jewish and Islamic rationalist tradition. Fraenkel is particularly interested in the impact of the Muslim philosopher Averrores, mediated through the Jewish Renaissance Averroist, Elijah Delmedigo, on Spinoza's philosophical-religious project. Finally, Alison Laywine's paper examines the early modern discussion of music theory—one of the traditional branches of "mixed mathematics"—and shows how this discussion is related to the emergence of a new branch of mixed mathematics: Galilei's mechanics. She argues that the research program underlying both disciplines makes use of mathematical reasoning and sense data as complementary criteria of truth and can ultimately be traced back to the music theory of the Alexandrian astronomer Claudius Ptolemy in the second century CE.

In the second section, *Creating Traditions*, Roger Ariew looks at the effort on the part of Descartes's disciples to create, from what little material Descartes left to them, a system of Cartesian ethics. Ariew contrasts the ethics of the 1685 Cartesian manual *Ethica* with the treatment given to it by three main Cartesian textbook writers: Pierre Sylvain Régis, Jacques Du Roure, and Antoine Le Grand. In these authors, we see how Descartes is made to lend authority in an area of thought in which we do not normally think of him today as having had much interest or influence. Next, François Duchesneau's paper gives us a picture of the emergence of a tradition of Leibnizianism in the generation of French natural scientists of the early eighteenth century, including Bourguet, who came to see the German philosopher as an important theorist of the phenomena of the living world.

The papers in the third section, *Rethinking Spinoza*, all call into question some aspect of the commonly received interpretation of this philosopher. Hasana Sharp rejects the view, held by many Spinoza scholars, that Spinoza's account of the highest kind of knowledge, the *scientia intuitiva*, is "incomprehensible". She also resists the widespread "elitist" interpretation of this kind of knowledge. Sharp argues that, on the one hand, *scientia intuitiva* is to some extent accessible to everyone. On the other—and this is a point usually not taken into account by scholars—its difficulty stems from the fragility of knowledge for all human beings *qua* finite beings, not just from the intellectual limitations of the *vulgus*. Syliane Malinowski-Charles undertakes to clarify the ontological status of inadequate ideas and passive affects in Spinoza by questioning the identity of the subjects of which Spinoza speaks when referring to the subjective and objective ideas of a mode. Against the widespread view which holds that for Spinoza, inadequate ideas and passive affects are "nothing," she argues that they must have a share in *Deus sive Natura* and proposes a new solution to the apparent discrepancy between these two selves. Finally, Yitzhak Melamed rejects the common perception of Spinoza as one of the founders of the Enlightenment and its humanistic world-view, arguing that Spinoza's position is best described as "anti-humanist". For Spinoza, Melamed shows, human beings are an insignificant part of the universe, their anthropomorphic conceptions of God and nature are baseless, and nothing sets them radically apart from other beings in nature. Even their moral coordinates do not refer to real features of things.

The three papers in the final section, *Legacies of Rationalism*, each attempt to pinpoint one of the core problems that served as a driving force behind the philosophical systems of Rationalist thinkers, or to show where we need to look in order to discern the core feature defining Rationalist thought. Steven Nadler's paper makes the case that, for a number of seventeenth-century philosophers, the problem of theodicy was the most pressing issue that exercised their minds and led them to develop their philosophical systems. Brandon Look's paper on the principle of sufficient reason in Leibniz's philosophy and beyond reminds us of the enduring importance of one of the central tenets of Leibniz's metaphysics. Ohad Nachtomy's paper considers another fundamental metaphysical problem in Leibniz: the related concepts of infinity and being. Nachtomy argues that, for Leibniz, both a nonactive law or program of action as well as a source of action or primitive force are required in order for there to be an actual being.

References

Ariew, Roger. (1998). *Descartes and the last scholastics*. Cornell: Cornell University Press.

Bennett, Jonathan. (2001). *Learning from six philosophers: Descartes, Spinoza, Leibniz, Locke, Berkeley, Hume*. Vol. 1. Oxford: Oxford University Press.

BonJour, Laurence. (1998). *In defense of pure reason: A Rationalist account of* a priori *justification*. Cambridge: Cambridge University Press.

Cohen, Claudine and Andre Wakefield. (2008). *Introduction to Protogaea, Gottfried Wilhelm Leibniz*. Chicago: University of Chicago Press.
Cottingham, John. (1988). *The Rationalists*. Oxford: Oxford University Press.
Des Chene, Dennis. (2000). *Physiologia: Natural philosophy in late Aristotelian and Cartesian thought*. Cornell: Cornell University Press.
Elster, Jon. (1975). *Leibniz et la formation de l'esprit capitaliste*. Paris: Aubier Montaigne.
Garber, Daniel. (1992). *Descartes' metaphysical physics*. Chicago: University of Chicago Press.
Gennaro, Rocco J., and Charles Huenemann, eds. (2002). *New essays on the Rationalists*. Oxford: Oxford University Press.
Haakonssen, Knud. (2006). The history of eighteenth-century philosophy: History or philosophy? In *The Cambridge history of eighteenth-century philosophy*, ed. Knud Haakonssen, 3–26. Cambridge: Cambridge University Press.
Harvey, Warren Zeev. (1981). A portrait of Spinoza as a Maimonidean. *Journal of the History of Philosophy* 19 (1981): 151–72.
Huenemann, Charles. (2008). *Understanding rationalism*. Stocksfield: Acumen Press.
Menn, Stephen. (1998a). *Descartes and Augustine*. Cambridge: Cambridge University Press.
Menn, Stephen. (1998b). The intellectual setting. In *The Cambridge history of seventeenth-century philosophy*, ed. Daniel Garber and Michael Ayers, Vol. 1, 33–86. Cambridge: Cambridge University Press.
Perler, Dominik. (1998). *René Descartes*. Munich: Beck.
Phemister, Pauline. (2006). *The rationalists: Descartes, Spinoza and Leibniz*. Cambridge: Polity Press.
Roth, Leon. (1924). *Spinoza, Descartes, and Maimonides*. Oxford: Clarendon Press.
Schneewind, Jerome B. (2004). Teaching new histories of philosophy. Proceedings of a conference, University Center for Human Values, Princeton University, April 4–6, 2003. Princeton: University Center for Human Values, Princeton University.
Sorell, Tom and G. A. J. Rogers. (2005). *Analytic philosophy and history of philosophy*. Mind Association Occasional Series. Oxford: Clarendon Press.
Zarka, Yves Charles. (2001). *Comment écrire l'histoire de la philosophie? Quadrige* 358. Paris: Presses universitaires de France.

Part I
Continuities Between the Premodern and the Modern

Chapter 2
Descartes on Human Nature and the Human Good

Lisa Shapiro

It might well seem that Descartes has little to say about the human good. After all, he did not author a work devoted to ethics, and his scattered remarks in his published works—the *Discourse on Method*, the *Passions of the Soul*, and the preface to the French translation of the *Principles of Philosophy*—are not particularly systematic; and in the *Passions*, where we might expect a developed moral theory, Descartes expressly disavows the approach of the moralist to adopt that of a *physicien*, that is, of a natural philosopher.[1] Nonetheless, I do think that we can reconstruct Descartes's ethics from these scattered remarks and the more sustained treatment of moral philosophy in his correspondence with Princess Elisabeth of Bohemia and the letters to Queen Christina of Sweden.[2] Elsewhere I have argued that Descartes subscribes to a virtue ethics, and one heavily influenced by Stoicism.[3] I do not want to rehearse that discussion here. However, I do want to consider a question that emerges from that interpretation. While different varieties of virtue ethics conceive of the human good differently, all seem to share a basic structural feature: they tie the human good to human nature. I call this feature *structural eudaimonism*. I want to consider whether Descartes, in his virtue ethics, is also a structural eudaimonist. I begin by setting out just what I mean by "structural eudaimonism", how it is to be understood as distinct from eudaimonism, and why it is worth considering whether Descartes subscribes to it. I then explore the view suggested by the *Meditations*, that the human good consists in the good of the human mind, the pursuit of knowledge. This view is complicated, however, by the fact that, for Descartes, a human being is not simply a

L. Shapiro (✉)
Simon Fraser University, Burnaby, BC, Canada
e-mail: lshapiro@sfu.ca

[1] In his contribution to this volume, Roger Ariew provides historical contextual evidence in support of this view.

[2] See letters to Elisabeth of 4 August 1645 (4:236ff; 3:256ff), 18 August 1645 (4:271ff; 3:259ff), 1 September 1645 (4:281ff; 3:262ff), 15 September 1645 (4:290ff; 3:265ff), 6 October 1645 (4:304ff; 3:268ff). 3 November 1645 (4:330ff; 3:276ff), January 1646 (4:351ff; 3:281ff), and to Christina of 20 November 1647 (5:81ff; 3:324). For Elisabeth's side of the exchange see my (2007). Translations from this edition will be cited internally as "CED," followed by the page number.

[3] See my (2008).

C. Fraenkel et al. (eds.), *The Rationalists: Between Tradition and Innovation*,
The New Synthese Historical Library 65, DOI 10.1007/978-90-481-9385-1_2,

mind, but a union of mind and body. To this end, it would seem that the human body ought to contribute something to our proper function or good. In Section 2.3, I consider what sense we can make of any role the body might play in the human good, and in Section 2.4 I suggest that Descartes is best read as appropriating the Stoic notion of *oikeiosis*. However, it is not clear how a purely mental good can be reconciled with this sort of account. I conclude that despite his virtue ethics, Descartes is best not read as subscribing to any kind of eudaimonism, not even a very weak structural eudaimonism.

2.1 Eudaimonism and Structural Eudaimonism

In *Nichomachean Ethics* 1.7 Aristotle claims that eudaimonia is the highest good, and that which is "always desirable in itself and never because of something else" (NE 1097a35–b1). While virtue is necessary for eudaimonia, it is not identical with eudaimonia itself. The Greek "eudaimonia" is often translated as "happiness", and though there has been much discussion about just how to understand the sense of "happiness" here, we do not need to resolve this issue. However, if we focus simply on the core of the position—that happiness is the highest good—it is hard to claim that Descartes is a eudaimonist. In correspondence with Elisabeth he writes, unequivocally: "true happiness is not the sovereign good; but it presupposes it, and it is the contentment or satisfaction of the mind which comes from possessing it" (18 August 1645; 4:275; CED 103). This view is reaffirmed in his brief exchange with Queen Christina, and there is no suggestion that he modifies it in the *Passions*. While Descartes agrees with the Aristotelian position that virtue is necessary for happiness and not identical with it, for him, *virtue*, not contentment or happiness, is the highest good. As Descartes defines it, virtue is "a firm and constant resolution to carry out to the letter all the things one judges to be the best, and to employ all the powers of one's mind in finding out what these are" (To Christina, 20 November 1647; 5:83; 3:325).[4] That Descartes puts his position the way he does makes it seem that he is simply rejecting eudaimonism altogether. Indeed, he writes to Elisabeth that Aristotle's account of the highest good is "not useful to us" (18 August 1645; 4:276; CED 104). The account is "not useful" simply because it concerns "the most accomplished of all men." Descartes's rejection of eudaimonism, then, seems tied to its focus (or at least Descartes's view of its focus) on the extraordinary human being, the one who has realized "all the perfections of which human nature is capable" (18 August 1645; 4:276; CED 104). How then ought we to make sense of Descartes as a virtue ethicist?

We can address this question by considering another central element of the Aristotelian eudaimonism, contained in what is often called the Function Argument.[5] According to this argument, each thing has a natural end or function,

[4]See also letter to Elisabeth 4, August 1645; 4:265; CED 97ff and PA a.153.

[5]The Function Argument is to be found at NE 1097b22-1098a20.

and the excellence—or highest good—of a thing is measured by how well it fulfills its function. The function of a knife is to cut, and so a knife is a good knife insofar as it cuts well. Similarly, human beings are good insofar as we fulfill the function of a human being; and we are excellent, that is we achieve our highest good, just insofar as we fully fulfill that function. For Aristotle, what is distinctively human is reason and so "the human function is the soul's activity that expresses reason" (NE 1098a6). Achieving the human good involves expressing reason well. In this paper, I am not concerned with the substance of the eudaimonist account, but rather with this particular structural feature of it. For Aristotle, the highest human good is tied to human nature. I refer to this feature as structural eudaimonism.

I will consider whether Descartes's ethics shares this structural eudaimonism, whether, for him, the human good is tied to human nature, and our fulfillment of that nature. While it might seem that a consideration of this point would consider whether Cartesian virtue derives from his account of human nature, this will not be my direct line of approach. Rather, my focus is on whether and how a notion of the human good can derive from the Cartesian account of human nature as a union of mind and body. In my concluding remarks, I will have something to say about Descartes's rejection of the perfectionism he takes as intrinsic to Aristotelian eudaimonism.

Let me make two preliminary points. First, it ought not be taken for granted that an early modern philosopher would be a eudaimonist in even this structural sense. We need only look at Descartes's contemporary and the author of the Third Objections, Thomas Hobbes, to see that this is so. Hobbesian human beings hold a natural right to self-preservation, but there is no human good tied to this natural right. As a consequence, its exercise leads naturally to a world where life is "solitary, poor, nasty, brutish and short."[6] Morality, which Hobbes casts as "right and wrong, justice and injustice" has "no place" in the state of nature. It is not tied to human nature or a distinctive human good, but rather emerges with the establishment of a social order.[7] If Descartes, who rejects some basic tenets of Aristotelian metaphysics, were to retain this structural feature of Aristotelian ethics it would be interesting just as a point of fact.

It would also be interesting in its substance. Within the Aristotelian framework, the structural eudaimonism on which I am focused is cast in functional terms. That is, it is grounded on there being an end, a final cause, proper to the human being, and this end is grounded in the form of the human being. But Descartes, in a direct rejection of Aristotelianism, denies that there are formal causes and final causes. To say that the human being has a proper function would seem to make no sense for Descartes. In considering whether Descartes subscribes to a form of structural eudaimonism, I do not want to go so far as to claim that there is a formal cause of the

---

[6]Thomas Hobbes, *Leviathan*, I.13.9

[7]See Hobbes, *Leviathan*: "Justice and injustice are none of the faculties, neither of the body nor mind. If they were, they might be in a man that were alone in the world, as well as his senses and passions. They are qualities that relate to men in society, not in solitude" (I.13.13).

Cartesian human being. I do, however, want to explore to just what extent a notion of an end or proper function can be found in Descartes's philosophical commitments.

2.2 The Meditations: The Nature of the Human Mind and the Human Good

From a consideration of the position Descartes articulates in the *Meditations*, it can certainly seem as if Descartes does subscribe to structural eudaimonism, and, in fact, not unlike Aristotle, takes the human good to consist in the good of the human mind. Aristotle, as we have seen, identifies the highest human good with the happiness that comes from excellence of rational activity, and expressly brackets the virtues of those capacities we share with plants and animals (nutrition and growth, and with animals alone, self-motion) as having "no part" in distinctively human virtue, and so no part in eudaimonia. This position seems consistent with Descartes's account of human nature in many respects. For Descartes, just as for Aristotle, the capacity for reason distinguishes humans from other animals. We have minds. Moreover, it is natural to slide from the meditator's claim in the context of radical doubt that "I am a thing that thinks," a thing whose essence is thinking, to one about human nature: human beings are just things that think, things whose essence is thinking. If this move is warranted, then a Cartesian structural eudaimonism is easily articulated.

Recall that the core of what I am calling structural eudaimonism is a function argument. And a function argument hinges on the thing at issue having a proper function, or a proper end that defines it as the thing it is. In the Aristotelian system, the function of a thing is given by its form, and though within the Cartesian framework it does not make sense to claim that the mind is a form in the Aristotelian sense, it still seems that Descartes takes the mind to contain its own end. The argument of the *Meditations* is meant to demonstrate that a thinking thing in itself not only strives for knowledge but also contains in itself the standard of truth and falsity. Defending this claim fully is well beyond the scope of this paper, but let me very briefly defend its plausibility. It should be clear that the *Meditations* itself is premised on the desire for knowledge. Not only does the meditator motivate his initial skeptical arguments from his desire to "establish anything at all in the sciences that was stable and likely to last," but this same desire propels him forward from the desperation of that skepticism, to continue to search "for something certain, or if nothing else, until I at least recognize for certain that there is no certainty" (7:24; 2:16). The drive for knowledge is intrinsic to the meditator and not foisted upon him from outside. His reflection on this drive for knowledge reveals it to entail the recognition that "I am a thinking thing." Equally, further reflection on what it is to be a thinking thing yields the criterion of certainty and knowledge. I take this to be the upshot of the Fourth Meditation. That meditation yields a method for avoiding error in our judgments of truth and falsity. That method, however, is premised on features of our nature as thinking things: that we have an idea of God and from that fact are able to assure ourselves that God is not a deceiver; and that we are able to perceive clearly and distinctly and are determined to affirm what we so perceive. Together these features of our nature entail that what we clearly and distinctly perceive is true. So long as we properly attend to our nature, and to our thoughts, we can avoid error by affirming only those ideas we perceive clearly and distinctly. A close reading of the end of the Fourth Meditation shows that it is very much a point about our nature as thinking things. That nature yields in itself a measure of truth and falsity. The mind then contains its own end—knowledge—that drives its activity. It thus makes sense to say that the mind has a function, the fulfillment of which defines the excellence of the mind.

If human nature is identified with the nature of mind, as the stylistic features of the *Meditations* can suggest, then it would seem that we are entitled to attribute a structural eudaimonism to Descartes. We only would need to spell out in more detail how what Descartes writes about the nature of mind in the *Meditations* translates into his account of virtue as the highest human good. This project does not seem to be particularly daunting, as there are resonances between Descartes's definition of virtue, as resolving to do what we judge to be the best, and the centrality of using our will well to the method for avoiding error.[8]

Even though this sort of reading is tempting, I do not think it can be Descartes's considered view. Given Descartes's dualist commitments, if our good derived from the good of the mind, then Descartes would need to treat the body's condition as merely an accidental good: though it might well contribute to our ability to be virtuous, it would not be necessary for it. However, it does not seem that Descartes takes bodily health in this way. When pressed by Elisabeth in correspondence,[9] Descartes concedes that a certain level of bodily health is essential to the well-functioning of our rational faculties. He writes:

> when I spoke of a true happiness which depends entirely on our free will and which all men can acquire without any assistance from elsewhere, you note quite rightly that there are illnesses which, taking away the power of reasoning, also take away that of enjoying the satisfaction of a rational mind. This shows me that what I have said generally about all men should only be extended to those who have free use of their reason and with that know the path necessary to take to reach this true happiness. (1 September 1645; 4:281f; CED 106f)

Descartes here suggests that our "power of reasoning" depends on our bodily health. Insofar as virtue requires a firm resolution to do what we judge to be the best, it depends on this power of reasoning. Thus, virtue, for Descartes, our highest good, depends on our bodily health. Moreover, it is notable that the passion of generosity is the "key to all the virtues" for Descartes. This claim is not surprising given Descartes's definition of generosity as the knowledge that one has a free

---

[8]For some discussion of this parallel, see my (1999).

[9]Elisabeth writes: "There are diseases that destroy altogether the power of reasoning and by consequence that of enjoying a satisfaction of reason. There are others that diminish the force of reason and prevent one from following the maxims that good sense would have forged and that make the most moderate man allow himself to be carried away by his passions and less capable of disentangling himself from the accidents of fortune requiring a prompt resolution" (4:269; CED 100).

will and the resolution to use that will well. The definition reflects his definition of virtue. What is surprising is that Descartes counts generosity as a passion, and as such dependent on a bodily state. Again the suggestion is that virtue is tied to the body in an essential way.

Nonetheless, that the human body should figure in Descartes's account of virtue ought not be startling. For despite the rhetorical devices of the *Meditations*, Descartes does not think the essence of the human being consists simply in the human mind. A human being for Descartes is a *union* of mind and body, after all. If Descartes does hold a structural eudaimonist view then the nature of the union, both mind and body, ought to figure in the account of his human good. The question is what role there is for the body to play in any account of the human good for Descartes.

2.3 A Bodily Contribution to the Human Good?

In the Sixth Meditation Descartes takes pains to distinguish two senses of nature. One, the proper sense, refers to the way in which bodies in motion are governed by the laws of nature. The other, an "extraneous label", imports a purpose extrinsic to the particular body at issue. The mechanism of a clock cannot but follow the laws of nature, no matter what time it tells. Claiming that the clock is not working properly—that it is not a good clock—involves introducing the extrinsic purpose of keeping time. The point applies generally to all bodies, and since the human body is simply another body, for Descartes, there should be nothing about the mechanism of the human body that can ground any good proper to it. No matter what, the body follows the laws of nature. To identify a function or end proper to it, it would seem we must appeal to some purpose extrinsic to it. The obvious candidate here would be some purpose the soul has: a body would thus be functioning well or not relative to the degree to which its condition serves the soul's ends. It thus seems that the human body, in and of itself, can offer nothing to any account of the human good.

Despite the caution of the Sixth Meditation, I do think that Descartes avails himself of a notion of an end intrinsic to the body itself. Through this intrinsic end it makes sense to talk of bodily well-functioning. I will call this a notion of the bodily good. To begin to see that Descartes ought to have such a notion of bodily good, consider how the soul, an entity really distinct from body, could arrive at a set of purposes for which it might put the body to use independently of a set of facts about how the body works. It does not seem that it could. Rather, it would seem that facts about the workings of the body provide constraints on the soul's aims insofar as it is embodied. Descartes's remark to Elisabeth that

> what makes it the case that, for example, we do not at all desire to have more arms, or better, to have more tongues, than we have, but that we do desire to be in better health or to have more riches, is only that we imagine that these latter things can be acquired by our conduct, or even that they are due to our nature, and that the same is not true of the others (4 August 1645; 4:266; CED 98)

seems to suggest as much. His point is that our desires and our purposes are shaped by certain natural facts about our bodies—the number of arms and tongues that we have—and we would do well to shape our other desires similarly, as constrained by facts about nature rather than as up to us. The workings of the body would thus not be good or bad relative to the soul's purposes, but rather a precondition of the soul's forming the goals it does.

In addition, Descartes does claim repeatedly that bodies function better or worse insofar as they are able to preserve themselves. This is implicit in his account of the difference between a living and a dead body:

> And let us judge that the body of a living man differs from that of a dead man as much as a watch or other automaton (that is, other self-moving machine) when it is wound and contains the bodily principle of the movements for which it is constructed, along with everything required for its action, differs from the same watch or other machine when it is broken and the principle of its movement ceases to act (PA a.6; 11:330–31)

Descartes here highlights an aspect of mechanism left in the background of his discussion in the Sixth Meditation. Machines like watches or clocks are self-moving; their design enables them to preserve themselves in motion. A properly constructed watch will, when wound, keep ticking. This self-preserving ability is a necessary condition for the machine's serving its purpose, and indeed any purpose at all. A watch that explodes when its hands point to twelve is not much of a time keeper, and it's not clear what it would be good for, except maybe a gag gift. What is true of a watch is true of the human body. The human body too is a self-moving mechanism. It has a bodily principle of movement, one that is *intrinsic* to the construction of the body, and this internal bodily principle allows the human body to be self-preserving. It keeps the workings of the body intact and ticking along, just as the mechanism of a watch keeps it intact and ticking away. This internal stability is essential to the soul's forming of embodied desires.

I want to claim that this internal stability, this tendency to self-preservation, is the bodily good. Consider an important disanalogy between the clock and the human body. The clock is an artifact whose internal principle of motion is given to it by the craftsman who made it and its user. While Descartes does draw an analogy between God and a craftsman in the Third Meditation, there God is crafting a mind with an idea of God. The human body is a natural object, and while Descartes does take God to introduce motion into nature, he does not take it that God animates each particular body individually with the touch of his finger. God provides the natural world as a whole with an initial push. Natural bodies come to have *their* mechanical composition simply by the movement of the parts of matter in accord with the laws of nature. The principle of movement driving our bodies is thus not dependent on any immediate external intervention, nor on any purposes driving their construction; there are no purposes driving their construction. The bodily principle allowing the human body to preserve itself is intrinsic; it is simply a function of its composition. It is worth noting that Descartes applies this idea to animals as well as to human bodies. He writes, "animals that lack all reason direct their lives entirely by bodily movements" (PA a.138, 11:431). Those bodily movements are internal principles of motion serving to preserve the body, to keep it intact. Since animals do not have souls for Descartes, we cannot say that their ability to preserve themselves serves a soul's purpose.

But why should we count this intrinsic capacity for self-preservation a good? Descartes claims that a body which preserves itself in this way is "more perfect" than one that does not. I suspect that perfect here is just a measure of the ability to remain in existence. A body continues in existence as the same body just insofar as it can preserve itself. Its organization or mechanical composition allows for this. If it loses its mechanical composition—or decomposes—it loses its ability to preserve itself and goes out of existence as that particular piece of organized matter—that body. In this way, it makes sense to speak of a bodily good. A bodily good for Descartes is just this intrinsic mechanical ability of a body to preserve itself and so remain in existence.

Even if we allow Descartes this notion of a bodily good, understood as an ability of self-preservation, we still find ourselves with a pressing question: How can this bodily ability figure in Descartes's account of the human good, and in particular in his claim that the highest human good is the firm and constant resolution to do what we judge to be the best?

2.4 *Stoic* Oikeiosis *and Descartes's Account of the Human Good*

To answer this question, I suggest that we look to the Stoic notion of *oikeiosis*. There are good historical reasons for taking Descartes to be influenced by Stoic ideals. First, it is clear that Descartes had some familiarity with Seneca, as he and Princess Elisabeth of Bohemia corresponded extensively regarding *De Vita beata*. Seneca's letters were certainly available, and he might well have consulted them. In addition, editions of Stobaeus's anthology of remarks (which contains the Hierocles passage) were published in the late sixteenth and early seventeenth century. One appears in 1575, and then there are others in 1609 and 1625. Judging from the number of extant copies in libraries, the work seems to have been in wide circulation. Also, an edition of other of Stobaeus' writings by Hugo Grotius was published in 1623. Grotius was a proponent of Stoicism and appropriated Stoic thought in the development of his political philosophy. There is thus good reason to think Stoic writings were widely available and much discussed. There are also textual parallels between Descartes's writings and those of the Stoics that make the case for influence quite compelling. However, after a consideration of the Stoic account we will be in a better position to see them and to understand the import of the Stoic influence for understanding Descartes's account of the human good I thus turn now to the Stoics.

The ground of Stoic ethics is widely recognized to lie in the notion of *oikeiosis*. There is much debate among classicists and scholars of ancient philosophy about just how to understand this notion, but to see how it can facilitate an understanding Descartes's notion of the human good, we need only have a general sense of the concept without settling on finer points of interpretation. And, though I will argue that it is very likely that Descartes was familiar with and influenced by this aspect of Stoic thought, it is less likely that his understanding was particularly refined.

Oikeiosis has been variously translated as "appropriation", "affinity", and "familiarization". All these translations share the connotation of "belonging to", and there is broad consensus that *oikeiosis* is a process through which individuals makes something their own—it is a process of appropriating something—of literally making it proper to oneself—or equally, of developing an affinity for things. (The standard Latin translation of the Greek is "conciliatio" which has a similar connotation.)

For the Stoics this process essentially involves a developmental story. Seneca's Letter 121 (to Lucilius) nicely expresses this commitment:

> There is a constitution for every stage of life, one for a baby, and another for a boy, another for a teenager, another for the old man. Everyone is attached [*conciliatur*] to the constitution he is in. A baby has no teeth—it is attached to this constitution, which is its own [*huic consitutioni suae conciliatur*]. Teeth emerge—it is attached to this constitution. (Seneca 2007, Letter 121.15)

The idea here is that human life is characterized by different stages, and at different stages of life, different things are appropriate to a human being.

We can use this passage from Seneca as a starting point for further understanding the Stoic notion of appropriation. Seneca identifies what is *oikeion* (or *conciliata*) as our bodily constitution, and there seems to be agreement among the Stoics that the first thing that is *oikeion* is our own body and its constitution. It is key to note that claiming that our bodily constitution is *oikeion* is not simply a matter of claiming that we *have* a body with a certain sort of constitution. Being *oikeion*, or appropriate, is something more active. To claim that our body is *oikeion* is to say that *we take* our body to belong to us and so to be constitutive of the individual we are. Identifying ourselves in this way in turn shapes our motivations to act, or what the Stoics term impulses. This brings out a third point about *oikeiosis*. Taking something to belong to us involves our caring about that thing. It is this connotation of caring that warrants translating *oikeiosis* as "affinity". This aspect of concern explains how our impulses are affected by what is *oikeion*, for the care through which we take something as our own is what moves us to act. The Latin translation of *oikeion* brings this point out. *Conciliatio* not only connotes a bringing together, but also suggests a uniting of interests and the causing of good will, and so implies a shift in inclinations to act. Tad Brennan sums up this point: "what it means to take something to be *oikeion* is that one treats it as an object of concern. In particular, when you think of something as *oikeion*, you think of its welfare as giving you reasons to act" (Brennan 2005, 158). So, insofar as we first take our bodies to belong to us, our first motivations or impulses are just to preserve our bodily constitution, to preserve ourselves. The centrality of self-preservation is laid out by Diogenes Laertius, who attributes it to Chrysippus:

> An animal has self-preservation as the object of its first impulse, since nature from the beginning appropriates it...The first thing appropriate to every animal, he [Chryssipus] says, is its own constitution and the consciousness of this...in constituting the animal, nature

appropriated it to itself. This is why the animal rejects what is harmful and accepts what is appropriate. (Diogenes Laertius 7.85-6 [SVF 3.178], Long and Sedley 1987, 346)

This passage highlights a basic element of the account, and at the same time it sets up a further move. The impulse to self-preservation is not something unique to human beings; we share this interest in self-preservation with animals, and indeed it is part of our nature to be animals. Moreover, animals develop just as do human beings. As we (and animals) mature, what we take to be our own—what we take ourselves to be—changes. We saw this position articulated in Seneca.

However, for the Stoics, human beings differ from other animals in having a rational faculty. As we develop this rational faculty, what is appropriate to us changes along another dimension as well. It is not that we cease to take our body as appropriate, but that the scope of what we appropriate or attach ourselves to enlarges. In the course of our development, we come to see ourselves as parts of successively larger and larger wholes. A passage from Hierocles, found in Stobaeus's anthology of remarks of ancient philosophers, summarizes the developmental story:

Each one of us is as it were entirely encompassed by many circles, some smaller, others larger, the latter enclosing the former on the basis of their different and unequal dispositions relative to each other. The first and closest circle is the one which a person has drawn as though around a centre, his own mind. This circle encloses the body and anything taken for the sake of the body. For it is virtually the smallest circle, and almost touches the centre itself. Next, the second one further removed from the centre but enclosing the first circle; this contains parents, siblings, wife, and children. The third one has in it uncles and aunts, grandparents, nephews, nieces, and cousins. The next circle includes the other relatives, and this is followed by the circle of local residents, then the circle of fellow tribesmen, next that of fellow-citizens, and then in same way the circle of people from neighbouring towns, and the circle of fellow-countrymen. The outermost and largest circle, which encompasses all the rest, is that of the whole human race. Once these have all been surveyed, it is the task of a well tempered man, in his proper treatment of each group, to draw the circles together somehow towards the centre, and to keep zealously transferring those from the enclosing circles into the enclosed ones. ... (Hierocles [Stobaeus 4.671, 1–673, 11], Long and Sedley 1987, 349)

According to Hierocles, initially we are concerned with our body, but as rational agents, we soon come to understand ourselves as also minded. These concerns form the centre of our sense of ourselves as individuals, and so lead us to act in such a way as to preserve our bodies. As we continue to develop, we come to situate ourselves as parts of successively larger wholes. Situating ourselves as parts of these wholes in turn shapes our motivations and reasons for actions: our sphere of concern enlarges successively, and we are moved to act for the sake of our family, our town, and the society of which we are a part. The task of a fully rational agent is to recognize the proper relation of part to whole. Recognizing her relation to other persons and things in the world in turn affects her motivations to act. She will see how her own individual interests are a function of the interests of the larger whole.

It should be clear that, for the Stoics, human nature contains the human good. The Stoics have a function argument. The function of a human being is just to live in accord with nature. We do this by striving to preserve ourselves. Human efforts of self-preservation are rational. They involve a conscious understanding of ourselves in relation to others. Living in accord with nature involves having a complete understanding—we fulfill our nature as human beings by properly cognizing how we fit into the world.

While there is much more to be said about the Stoic account of *oikeiosis* and right action, this is enough for us to examine textual parallels between Descartes and the Stoics. In the next section I will consider the how this Stoic influence can help us in thinking about Descartes's conception of the human good.

It should be clear from the earlier discussion of Cartesian bodily good that there is a parallel between Descartes's and the Stoics' concern with self-preservation. It should be noted too that neither Descartes nor the Stoics explain the tendency to self-preservation by the action of a formal cause. For both, self-preservation is a matter of the organization of matter. But there are other textual similarities as well. Compare this remark Descartes makes to Elisabeth in correspondence to Hierocles's remark about the circles of concern:

> even though each of us is a person separate from others and, by consequence, whose interests are in some manner distinct from those of the rest of the world, one must, all the same, think that one does not know how to subsist alone and that one is, in effect, one part of the universe and, more particularly even, one part of this earth, one part of this state, and this society, and this family, to which one is joined by his home, by his oath, by his birth. It is always necessary to prefer the interests of the whole, of which one is a part, to those of one's person in particular, though with measure and discretion. (15 September 1645; 4:293, CED 112.)

The resonances between the two are easy to hear. Like Hierocles, Descartes takes us to first consider ourselves as individuals. Moreover, for both, our development consists in our taking ourselves to be parts of successively larger wholes, and our understanding of ourselves as parts of those wholes shapes our reasons to act.

But with the Stoic background in mind, other passages also stand out as having a neo-Stoic cast to them. Towards the beginning of the Sixth Meditation, in reviewing all his previous sensory beliefs, the meditator singles out one in particular: "my belief that this body, more than any other, belonged to me had some justification" (7:76; 2:52). The language of "belonging" here resonates with that of *oikeiosis.* And the way this notion of belonging is spelled out does as well. Our bodies are properly our own, for Descartes, in that we experience their pains and pleasures, and not those of others. This fact, moreover, gives us impulses to avoid or pursue those pains and pleasures. Descartes's account of the mind-body union is meant to explain these facts in more detail. While the explanation offered in the Sixth Meditation is not particularly telling with respect to potential Stoic influence, a discussion in the *Passions* aa.107–111 of our first passions, those we felt when the soul "was originally joined to our body," is more so. There Descartes writes:

> For it seems to me that our soul's first passions, when it was originally joined to our body, must have been due to the blood, or other juice entering the heart, sometimes being a more suitable nourishment than the usual for maintaining the heat in it which is the principle of life. That caused the soul to join this nourishment to itself in volition, that is, to love it... (PA a.107, 11:407)

According to this somewhat bizarre natural history, the soul in effect appropriates the body. It takes the body as its own—joins it to itself in volition—and in doing so comes to care about the state of the body. We get a similar story about the soul's disaffection with the body's constitution in its first feeling of hate. Similar stories of the soul's attitude towards the body's constitution explain joy, sadness and desire. The discussion is remarkably odd, but begins to make some sense from the point of view of the Stoic account of *oikeiosis*. Indeed, while Descartes does not advert to the Stoics here, I cannot imagine that he did not have the Stoic doctrine of *oikeiosis* in mind, for precisely what is detailed is the attachment of the soul to our bodily constitution. The soul, upon being joined with the body, takes it as belonging to it, and has concern for it, and this concern plays a causal role in the thoughts—the passions it feels—and its dispositions to act. I do think the sum of the textual coincidences are enough to lend plausibility to the idea that Descartes's account of human nature was informed by that of the Stoics.

2.5 Descartes, Human Nature, and the Human Good

We can now consider how this Stoic aspect of Descartes's account of human nature contributes to our consideration of Descartes's structural eudaimonism, that is, of the role our bodily nature might play in his account of the human good, and of whether he is afforded an account of the human good as tied to our nature as a *union* of mind and body. In this regard, there is a crucial difference between Descartes's philosophical position and that of the Stoics. For the Stoics, the development of our rational faculties is part of our natural biological development. There are many questions about how this story is supposed to go, but it is clear that the Stoics are materialists through and through. For Descartes, the mind is really distinct from the body, even while the two are joined together. Insofar as Descartes takes mind as really distinct from body, it does not seem that he can claim that our rational faculties emerge from the motions of those bodies making up our body. So, even if we do take Descartes as espousing an *oikeiosis*-like doctrine about our bodily good, that does not help us to understand how his conception of the human good is tied to his conception of human nature. That is, it does not help us to understand how he might subscribe to a form of structural eudaimonism. While we might be able to make sense of a part of the human good that is tied to the nature of mind and another part that is tied to the nature of body, we are not yet afforded an account of how these two parts fit together. How are we to integrate these two aspects of the human good?

In part to answer this question, but also to tie things up, let me consider how Descartes's claim that *virtue* is the sovereign good fits into this account. Recall that virtue, for Descartes, is simply having a firm and constant resolution to do that which we judge to be the best. As we have seen, it can be tempting to take this account of virtue as emerging from Descartes's account of the nature of mind. His definition of virtue lends primacy to the will, through its firm resolution, and on the faculty of judgment. Both are squarely situated in the mental for Descartes. Nonetheless, I have argued that for Descartes, the human good is also squarely rooted in our bodily constitution. How does the role of our bodily constitution in the account of the human good get us Cartesian virtue as the highest human good?

Let me first note a peculiarity of Descartes's account of virtue. For Descartes, virtue requires only that we are resolved to do what we judge to be the best. This seems quite weak as an account of virtue. After all, what we judge to be the best may not, in fact, be the best course of action. Our practical judgments are often wrong. The account can be strengthened a bit by remarking that in judging a course of action to be the best not just any judgment will do. Rather, Descartes might well insist, and it seems he does, that virtue consists in resolving to act on our *best* judgments about what is best. But even this is a weak account of virtue, for even our best judgments can be wrong. Thus, according to Descartes we can be both virtuous and mistaken in choosing our course of action.

Here Descartes clearly parts with the Stoic account, as for the Stoics virtue is an all or nothing affair. In fact, it is usually nothing. Since only the truly extraordinary sage will manage to arrive at virtue, where all her judgments are correct, most of us are left in a state of viciousness, trying but failing to get things completely right. For Descartes, the standard is not so high. We can all be virtuous if we try.

I want to suggest that Descartes's tempered account of virtue is actually grounded in his recognition that human beings are not simply minds, but minds united with bodies. The telling point here is the difference between the method for avoiding error and virtue. Like virtue, the method for avoiding error involves a resolution to judge the best we can. In that case, however, we are to judge the best we can regarding truth or falsity. Here, there is no room for error. We judge poorly when we get things wrong, when we mistake the false for the true and vice versa. Our best judgments must be correct in order for us to achieve our epistemic ends. Our moral judgments, however, need not meet this threshold. Why not? Descartes's account of human nature acknowledges that we are not supernatural creatures, but rather very much a part of nature, through our bodies. We thus find ourselves in the midst of a very complex world. Though we are constantly striving to understand our proper place in this world, we find ourselves without what Elisabeth terms "an infinite science."[10] Descartes's account of virtue recognizes that taking action requires making timely decisions based on incomplete information, and living with those decisions. His considered view is that we can achieve this by developing good habits of judgment. A quick glance at titles of his works—the *Rules for the Direction of the Mind, Discourse on the Method of Right Reason*—and even the method for avoiding error of the Fourth Meditation, shows Descartes to be very much concerned with the *process* of judging, perhaps more so than he is with the outcomes. It is hard to know just what judging well involves, but for our purposes here, we need only note that following the proper process is sufficient for virtue. Given that we are embodied,

[10] See letter to Descartes, 13 September 1645; 4:289; CED 110.

and so are subject to the constraints of time and imperfect knowledge, we can only judge the best the best we can.

While we can integrate the bodily and the mental aspects of the human good in this way, it is not clear to me whether we should hold out a hope of integrating them any further, to demonstrate that Descartes does subscribe to a coherent structural eudaimonism. As I have shown, there are strands of a structural eudaimonist position in Descartes's philosophy. Within the *Meditations*, we find a notion of a good of the mind, and in his ethical writings we can find an account similar to Stoic *oikeiosis*, suggestive of a bodily good. However, it is not clear that, given Descartes's dualism, these two accounts can be brought together. But even if we do manage to find a way of attributing to him a coherent structural eudaimonism, it does not seem that outside of the epistemic context Descartes espouses the sort of perfectionism we take to be essential to eudaimonism. Cartesian virtue does involve using our natural faculty of judgment well, and even excellently, but for him in the practical context even our best possible judgments seem likely to fall short simply given the fact that we are finite beings, with a limited amount of time to make decisions with imperfect knowledge. And though we do fall short, for Descartes, these errors of practical judgment do not preclude virtue. Given that Descartes does not think that virtue—the realization of the highest human good—requires our perfecting our nature, he does not need to see our good as tied to our nature. That is, insofar as he rejects the perfectionism characteristic of eudaimonism, he has no need for the structural eudaimonism from which eudaimonist accounts are derived.

References

Brennan, Tad. 2005. *The Stoic life: Emotions, duties, and fate*. New York: Oxford University Press.

Long, Anthony and David Sedley, eds. 1987. *The Hellenistic philosophers*. Vol 1. Cambridge: Cambridge University Press.

Seneca. 2007. *Seneca: Selected philosophical letters*. Ed. and trans. Brad Inwood. New York: Oxford University Press.

Shapiro, Lisa. 1999. *Cartesian generosity* In Norms and modes of thinking in Descartes. *Acta Philosophica Fennica* 64: 249–275.

Shapiro, Lisa. 2007. *The correspondence between Princess Elisabeth of Bohemia and René Descartes*. Chicago: University of Chicago Press.

Shapiro, Lisa. 2008. Descartes's ethics. In *A companion to Descartes*, ed. Janet Broughton and John Carriero, 445–464. Oxford: Blackwell.

Chapter 3
Spinoza on Philosophy and Religion: The Averroistic Sources

Carlos Fraenkel

Spinoza has often been recruited for genealogies of modernity.[1] Although these genealogies sometimes yield interesting insights, I doubt that "modernity" is a fruitful *philosophical* category on which Spinoza's relevance depends. One important aspect of Spinoza's work which is normally given a prominent place among his contributions to modernity is his critique of religion. This critique in my view has not yet been adequately understood within the context of his philosophical project. Clarifying its purpose, of course, depends on understanding the general role of religion and its relation to philosophy in Spinoza's work. While I surely will not be able to settle the matter in this paper, I would like to draw attention to what I take to be an important piece of the puzzle: the distinctly Averroistic aspects of Spinoza's approach to the relationship between philosophy and religion in the writings preceding the critique of religion set forth in the *Tractatus Theologico-Politicus*. The position of the *TTP* in a sense radicalizes the stance on philosophy and religion advocated by Averroes in his chief philosophical-theological work, the *Faṣl al-maqâl* (Decisive Treatise). It is highly likely that Spinoza was familiar with Averroes' main claims, for they are taken up by the Jewish Renaissance Averroist Elijah Delmedigo (d. 1493) in his Hebrew treatise *Behinat ha-dat* (The Examination of Religion), which was among the books in Spinoza's library. All parallels between Spinoza and Averroes that I will point out can be explained on the assumption that Spinoza read Delmedigo's treatise, and, as I will show, there is considerable evidence suggesting that he did. Let me stress, however, that my purpose is not to hunt for Spinoza's "sources." Rather, my assumption is that understanding the critical dialogue that Spinoza conducted with earlier and contemporary philosophers is indispensable for illuminating important features of his thought.

Before turning to the question of Spinoza's Averroism, let me briefly outline the underlying broader issues that provide the context for the present paper. My goal is

C. Fraenkel (✉)
Department of Philosophy, McGill University, Leacock Building, 855 Sherbrooke Street, West Montréal, Québec H3A 2T7, Canada
e-mail: carlos.fraenkel@mcgill.ca

[1] See e.g. Israel, who attempts to trace what he considers to be the distinctive features of modernity back to Spinoza (2001 and 2006). Compare also Goetschel (2004).

C. Fraenkel et al. (eds.), *The Rationalists: Between Tradition and Innovation*, The New Synthese Historical Library 65, DOI 10.1007/978-90-481-9385-1_3,

to explain a striking inconsistency in Spinoza's treatment of scripture by taking two intellectual contexts into account: the philosophical interpretation of the purpose of religion in medieval Islamic and Jewish thought and various approaches to scripture in the Netherlands of the seventeenth century. The problem I am interested in is this: In his critique of religion in the *TTP* Spinoza develops an exegetical method by which he intends to show that scripture contains no truth and, therefore, cannot interfere with philosophy.[2] Whereas philosophy determines what is true and false, religion based on scripture secures obedience to the law.[3] On the other hand, there is a significant number of passages throughout Spinoza's work—from the *Cogitata Metaphysica* to the *Ethics* and the late correspondence with Henry Oldenburg—in which he attributes a true core to scripture, often presented as its allegorical content. My main thesis is that this inconsistency is best explained by assuming that Spinoza is committed to two projects that he ultimately was unable to reconcile: he wants to use religion as a replacement of philosophy that provides the basis for the best life accessible to non-philosophers, and he wants to refute religion's claim to truth in order to defend what he calls "the freedom to philosophize."

The concept of religion as a replacement of philosophy which educates and guides non-philosophers is the standard view of medieval Islamic and Jewish philosophers that Spinoza knew well through his study of medieval Jewish philosophy. The main idea is that religion's content according to its literal sense—Biblical narratives, laws, rituals, and so forth—is a pedagogical-political program devised by philosophers to guide non-philosophers. The allegorical content of religion, on the other hand, corresponds to the doctrines demonstrated in philosophy. Religion's authority thus depends on the assumption that the teachings of religion are true on the allegorical level. Spinoza calls this position "dogmatic" and describes and rejects it in the *TTP*.[4] The "dogmatist," thus Spinoza's main criticism, instead of strictly separating philosophy from theology, turns theology into the "handmaid of philosophy" (*ancilla philosophiae*).[5]

The "dogmatic" approach was first introduced into medieval Islamic and Jewish philosophy by the medieval Muslim philosopher al-Fârâbî (d. 950).[6] According to al-Fârâbî, "religion" (*milla*) is an "imitation of philosophy" (*muḥâkîyya li-l-falsafa*) (*Taḥsîl al-sa'âda*, Ar. 185; Eng. 44).[7] Hence religion "comes after philosophy, in general, since it aims simply to instruct the multitude (*ta'alîm al-jumhûr*) in theoretical and practical matters that have been inferred in philosophy in such a way

[2] See in particular *TTP* 7.

[3] See in particular *TTP* 12–15.

[4] See Chap. 7 and 15.

[5] Cf. the title of *TTP* 15 (A 482; G iii. 180). I quote the *Tractatus Theologico-Politicus* in the new edition prepared by Fokke Akkerman (1999) (=A and page no.). I add references to Carl Gebhardt's edition (1925) (=G, volume no., and page no.), according to which I also quote all other writings of Spinoza.

[6] The following paragraph summarizes what I elaborated in Fraenkel (2008b).

[7] Al-Fârâbî's most elaborate discussion of religion is the *Kitâb al-milla* (*Book of Religion*).

as to enable the multitude to understand them by persuasion or imaginative representation, or both" (*Kitâb al-hurûf*, secs. 142–143). The difference between the philosopher and the prophet comes down to this: the prophet, in addition to intellectual perfection, also has the skills of an orator, poet, and legislator that allow him to translate philosophical insights into a language and a set of practical rules accessible to non-philosophers (the "multitude"). Religion is thus integrated into a philosophical framework as a pedagogical-political program for non-philosophers. In this sense Spinoza can say that religion is conceived as philosophy's "handmaid." One implication of this view is that a religious text, if understood literally, is similar but not identical to the philosophical doctrines it imitates. If understood as an allegorical representation, however, it can be translated, as it were, into these doctrines by means of allegorical interpretation. A standard example from the medieval Islamic and Jewish context is God's description as a king in scripture which is seen as a pedagogically useful metaphorical imitation of the philosophical doctrine of God occupying the first rank in the hierarchy of existents. The notion of the king conveys an approximate idea of God's rank to non-philosophers who cannot understand the ontological order, but who do understand the political order.[8] In other words: literally, the representation of God as a king is pedagogically and politically useful but not true; allegorically, on the other hand, it is true but not pedagogically and politically useful. The two most prominent "dogmatists" at the end of the early medieval period were the Muslim philosopher Averroes (d. 1198) and the Jewish philosopher Maimonides (d. 1204), who were also the last important representatives of the Aristotelian school in Muslim Spain. Both worked out an interpretation of Islam or Judaism respectively as a philosophical religion, on the basis of al-Fârâbî's model for conceiving the relationship between philosophy and religion. Spinoza in turn became first acquainted with the dogmatic approach by studying the work of Maimonides, as well as that of other medieval and Renaissance Jewish philosophers. In the *TTP* he uses Maimonides as an example to first illustrate the dogmatic approach and then reject it.

I have shown in detail elsewhere that before Spinoza started working on the *TTP* in 1665, he consistently endorsed the dogmatic position whenever he discussed the character of scripture.[9] In what follows, I will briefly summarize the conclusions of that paper insofar as they are relevant for understanding my present argument. Two examples will suffice to illustrate the different aspects of what I take to be Spinoza's early dogmatism. The first is a passage from *Cogitata Metaphysica* 2.8, in which Spinoza discusses God's will. The problem at stake is this: How are we to understand passages in scripture according to which "God hates some things and loves other things" since, taken literally, they imply that God's will is affected by and reacts to things he created and hence is mutable? This appears to contradict the view of the philosopher, according to which God's will is immutable:

---

[8]See e.g. *Taḥsîl al-sa'âda*, Ar. 185; Eng. 45, quoted by Averroes in his Commentary on Plato's *Republic*, 30. Cf. Maimonides, *Guide of the Perplexed*, 1.8–9.

[9]See Fraenkel (2008a).

> But when we say that God hates certain things and loves certain things, this is said in the same way as scripture says that the earth will spit out human beings and other things of this kind. That God, however, is not angry at anyone, nor loves things as the multitude [*vulgus*] believes, can be sufficiently derived from scripture itself. For this is in Isaiah and more clearly in Paul's Epistle to the Romans, chapter 9. [...] Finally, if in the holy scriptures some other things occur, which induce doubt, this is not the place to explain them; since here we only inquire into the things which we can grasp in the most certain way through natural reason [*ratione naturali*]; and it is sufficient that we demonstrate these clearly in order to know that scripture must also teach the same things [*ut sciamus Sacram paginam eadem etiam docere debere*]; because the truth does not contradict the truth [*veritas veritati non repugnat*] and scripture can not teach the absurdities [*nugas*] which the multitude imagines. [...] Let us not think for a moment that anything could be found in sacred scripture that would contradict the Natural light [*quod lumini naturae repugnet*]. (G i. 264–265)

The conflict between the philosophical doctrine of God's will and scripture is resolved in the way most medieval Muslim and Jewish rationalists would resolve it: the statements about God's love and hate in scripture must be understood allegorically. Only the *vulgus* understands them literally. Moreover, the correct understanding of God's love and hate can be found in scripture itself: in both the prophets of the Hebrew Bible (Isaiah) and in the New Testament (Paul). The criterion to determine which passages of scripture are to be understood literally and which allegorically clearly is their agreement or disagreement with the corresponding philosophical doctrine. Contradictions between philosophy and scripture derive from the fact that scripture does not teach things *more philosophico*, i.e., in the way we grasp them when we inquire into them by means of "natural reason." But since the truth arrived at by reason is the same as the truth contained in scripture, we can rest assured that nothing clearly demonstrated by reason contradicts what scripture teaches. The character of the teachings of scripture is adapted to the imagination of non-philosophers. Understood literally, they amount to absurdities, but the philosopher-exegete should in principle be able to make the philosophical content visible within the non-philosophical form. Spinoza in this passage clearly adopts the "dogmatic" position, attributed to Maimonides in the *TTP*.

The second passage is taken from Spinoza's first letter to Willem van Blyenbergh, written in January 1665. In his response to Blyenbergh's questions, Spinoza explains, among others, why, according to the Biblical account, God commanded Adam not to eat from the tree of knowledge, although, according to the philosopher, he determined his will to transgress the command:

> I say that scripture, because it is particularly adapted and useful to the multitude [*plebs*], always speaks in human fashion [*more humano*], for the multitude is unable to understand the higher things. For this reason I believe that all that God has revealed to the Prophets as necessary for salvation is set down in the form of laws [*legum modo*]. On this account the Prophets invented entire parables [*integras Parabolas Prophetae finxerunt*] representing God as a king and lawgiver, because he revealed the means [leading to] salvation and perdition and is their cause. The means, which are nothing but causes, they called laws and wrote them down in the form of laws. Salvation and perdition, which are nothing but effects necessarily resulting from these means, they described as reward and punishment, putting their words more in accordance with that parable than with the truth, constantly representing God as human, now angry, now merciful, now looking to what is to come, now jealous and suspicious, and even deceived by the devil. So philosophers and likewise all who have

> risen to a level beyond law, that is, all who pursue virtue not as a law but because they love it as something very precious, should not find such words a stumbling-block. Therefore the command given to Adam consisted solely in this, that God revealed to Adam that eating of that tree brought about death, in the same way that he also reveals to us through the natural intellect [*per naturalem intellectum*] that poison is deadly. (G iv. 92–94)

In this passage "revelation" refers to the prophet's knowledge of the means leading to salvation and perdition, of which God is the cause. This knowledge is comparable to the knowledge revealed to a biochemist by means of his natural intellect about the destructive effect of poison, of which God is equally the cause. If the prophet were to address a group of philosophers, he would explain the means leading to perdition and salvation *more geometrico*, in the same way as the bio-chemist would offer a causal explanation for the danger of poison if he were to address a group of scientists. But since the prophet's task is to instruct non-philosophers, he has to speak *more humano*—i.e., "in the language of human beings" as Spinoza puts it using a Maimonidean formula. For this purpose he composes a parable describing God as a king and lawgiver who issues commandments and prohibitions, who is pleased about obedience and angry about disobedience, and who rewards the former and punishes the latter.

Whereas from the passage in *Cogitata Metaphysica* 2.8 we learned that scripture's anthropomorphic representation of God has an allegorical sense, here we learn in which way the literal sense is useful to non-philosophers. By speaking of God *more humano* and translating causal connections into laws associated with rewards and punishments, scripture is able to replace for the non-philosopher philosophical insight as a guide to virtuous action. This I take to be the most important reason for adopting the dogmatic position: it allows preserving the authority of scripture as the basis of popular religion which provides a pedagogical-political program replacing philosophy for non-philosophers.

If until about 1665 Spinoza's position on the relationship between philosophy and scripture is indeed the same that he rejects as "dogmatism" in the *TTP*, i.e., the position, according to which theology is the *ancilla philosophiae*, the issue becomes more complicated after 1665 when he begins to work out his critique of religion, published in 1670 as part of the *TTP*. But despite the critique of religion in the *TTP*, different versions of the dogmatic position reappear throughout all of Spinoza's later writings, for the most part reflecting the medieval position of philosophers like al-Fârâbî, Maimonides, and Averroes. What all the passages in question have in common is this: none of them can be justified through the exegetical method that Spinoza promises to adopt in the *TTP*, namely "to neither affirm anything of [scripture] nor to admit anything as its doctrine which I did not most clearly derive from it."[10] To put it in a provocative way: If Spinoza had never written his critique of religion, these passages, together with those of his earliest writings, would have allowed him to claim that the allegorical content of scripture is never in conflict with what the *Ethics* teaches philosophers *more geometrico*, and that the literal content

[10] *TTP* Preface; Spinoza elaborates the method in *TTP* 7.

of scripture teaches non-philosophers *more humano*, i.e., by means of parables and laws, an imitation of the doctrines of the *Ethics*.

Finally I claim that the dogmatic position, which has philosophy determine the true core of religion, is not only compatible with the philosophical project in the *Ethics*, but also with the freedom to philosophize that Spinoza sets out to defend in the *TTP*. It is clear that Spinoza's main opponent in the *TTP* is not the dogmatic position, but the position of the Calvinist Church in the Netherlands of the seventeenth century, in particular the view that the authority of scripture overrides the authority of reason. Spinoza describes this position as "skepticism" in the *TTP* and contrasts it with the dogmatic position.[11] It is this form of "skepticism" that turns philosophy into the "handmaid of theology." This in turn is the chief threat to the *libertas philosophandi* according to Spinoza.[12]

Let me now address three important objections to my thesis concerning Spinoza's early dogmatism which can be met, I contend, if the Averroistic character of his dogmatism is recognized.[13] The first objection is that I am wrong to claim that until 1665 Spinoza consistently endorsed the dogmatic position, for there are three passages in his early writings in which he clearly states that philosophy and theology contradict each other. These are the *scholium* to *Principia Philosophiae Cartesianae* 2.13, *Cogitata Metaphysica* 2.12, and *Epistola* 23 to Blyenbergh. In the last of these passages the alleged contradiction is most clearly formulated:

> Furthermore, I should like it here to be noted that while we are speaking philosophically [*Philosophice loquimur*], we ought not to use the language of theology. For since theology has usually, and with good reason, represented God as a perfect man, it is therefore appropriate that in theology it is said that God desires something, that God is affected by anger through the deeds of the impious and delights in those of the pious. But in philosophy, where we clearly perceive that to ascribe to God those attributes which make a man perfect would be as wrong as to ascribe to a man the attributes that make perfect an elephant or an ass, these and similar words have no place, and we cannot use them without utterly confusing our concepts. So, speaking philosophically, we cannot say that God wants something from somebody, or that something angers or delights him. For these are all human attributes, which have no place in God. (G iv. 98)

The second objection is that Spinoza not only stresses the contradictions between philosophical and theological propositions, but also shows no interest in resolving them by allegorically commenting on scripture as Maimonides does in his chief philosophical-theological work, the *Guide of the Perplexed*.[14]

The third objection, finally, concerns my claim that the dogmatic position is consistent with the *libertas philosophandi* that Spinoza defends in the *TTP*. This seems

---

[11] See *TTP* 15.

[12] In the Preface to the *TTP*, Spinoza describes "skepticism" as the "one obstacle" that prevents potential philosophers from philosophizing (A 74; G iii. 12). Cf. *Epistola* 30.

[13] For a discussion of why Spinoza adopted the dogmatic position in his early writings, why he rejected it in the *TTP*, and why he continued making use of it even after dismissing it, see again Fraenkel (2008a).

[14] See the programmatic passages in *Guide* 1, Introduction and *Guide* 2.2.

to be contradicted by the fact that Spinoza criticizes Maimonides in the *TTP* for introducing a form of philosophical tyranny into scriptural exegesis. According to Spinoza, *libertas philosophandi* not only means that philosophers must be safe from persecution in the name of religion, but that every citizen has the right to believe whatever he or she thinks is right on the basis of scripture, no matter whether or not this belief corresponds to what has been demonstrated in philosophy. If Maimonides' view were correct, Spinoza writes,

> it would follow that the multitude, which for the most part does not know demonstrations or has no leisure for them, could admit of scripture only that which is derived from the authority and testimony of philosophers [*de Scriptura nihil nisi ex sola authoritate testimoniis philosophantum admittere poterit*], and would therefore have to assume that philosophers cannot err in their interpretations of scripture. This would indeed be a novel form of ecclesiastical authority, with very strange priests or pontiffs, more likely to excite the multitude's ridicule than veneration. (*TTP* 7 [A 316; G iii. 114])

All three objections can be met, I contend, once we recognize that Spinoza's dogmatism in important respects does not follow Maimonides but Averroes, with whose position he in all likelihood became familiar through Elijah Delmedigo. As I mentioned earlier, both Averroes and Maimonides use al-Fârâbî's model for conceiving the relationship between philosophy and religion to interpret their respective religious traditions as philosophical religions. There is, however, one difference that is crucial for my present purpose. Whereas for Averroes the true doctrines constituting the allegorical content of scripture must remain the exclusive domain of the philosophers who have the intellectual capacity to understand them, for Maimonides they can and must be made accessible at least partly to non-philosophers as well: through allegorical interpretation and religious legislation.[15] The importance of this difference between Maimonides and Averroes was already noted by Shlomo Pines. According to Pines, Maimonides was influenced by the ideology of the Almohads on this point, the North African Berbers who conquered Spain in the twelfth century and "compelled all their subjects to profess an official theology." This theology was derived from the system of the *mutakallimûn*, "who were the official theologians of the Almohad kingdom" (Pines 1963, cxviii–cxix).[16] Maimonides seems to have thought that all members of the religious community can be compelled to adopt true opinions—the doctrine of God's incorporeality for instance—through religious legislation. These true opinions must then be reconciled with scripture through allegorical interpretation.[17] In a sense, therefore, Maimonides represents a deviation from the standard version of the dogmatic position. But because of Maimonides' enormous impact on subsequent Jewish philosophy, his version of dogmatism was adopted by most Jewish philosophers from the thirteenth century to the Early Modern period. This explains why philosophical commentaries on Biblical books became one of the main genres of Jewish philosophy throughout this period.

[15] For the following paragraph, see the more detailed discussion in Fraenkel (2010).

[16] Cf. Stroumsa (2005).

[17] See in particular *Guide* 1.35.

From the point of view of an Averroist, however, Maimonides' project constitutes a problematic amalgamation of philosophy and theology, for Maimonides introduces philosophy into disciplines where for Averroes it is completely out of place: theology and jurisprudence. This is also the main criticism directed by Elijah Delmedigo against Maimonides. Like Averroes he stresses that philosophy and theology must be strictly kept apart. Let us briefly examine how Averroes argues for this separation in his chief philosophical-theological work, the *Faṣl al-maqâl wa-taqrîr ma bayn al-sharî'a wa-al-ḥikma min al-ittiṣâl* (Decisive Treatise and Determination of the Relationship between the Divine Law and Philosophy). In contrast to Latin Averroists, Averroes holds that no genuine contradiction between philosophy and religion can exist. Islam contains the truth and exhorts all Muslims to pursue it. The formula "*veritas veritati non repugnant*" that we saw in *Cogitata Metaphysica* 2.8 appears as follows in Averroes:

> Since this Law [*sharî'a*] is true and calls to the reflection leading to cognition of the truth, we, the Muslim community, know firmly that demonstrative reflection cannot lead to something differing with what is set down in the Law. For the truth does not contradict the truth [*al-ḥaqq lâ yuḍâdd al-ḥaqq*]; rather, it agrees with and bears witness to it. (*Faṣl al-maqâl*, 8–9)[18]

Averroes, of course, knows that this cannot be the case if the *sharî'a* is understood literally. For then it contains much that is at odds with what philosophy demonstrates. The reason for this is that for Averroes, like al-Fârâbî, there is an important "difference in human nature [*ikhtilâf fiṭra al-nâs*]," namely that between philosophers and non-philosophers, and that the divine Law is addressed to all Muslims, and not only to the philosophers among them (*Faṣl al-maqâl*, 10). To achieve this, the prophet proceeds as follows: for one thing he calls the philosophers to pursue true knowledge on the basis of demonstrations. In addition he translates this knowledge by means of rhetorical and dialectical arguments, as well as poetic representations, into a language accessible to non-philosophers. As a consequence, contradictions arise between the literal sense of the divine Law and the doctrines demonstrated by the philosophers. These contradictions can be solved, according to Averroes, through "exegesis" (*ta'wîl*) which discloses the "allegorical sense" (*bâṭin*) of the divine Law.[19] The decisive point for my present argument is that allegorical exegesis is permitted only to philosophers according to Averroes. The difference between philosophers and non-philosophers with respect to the truth is thus twofold: Only the philosophers have access to the truth through scientific demonstrations and only the philosophers have access to the "allegorical sense" of the divine Law. For Averroes pointing out in public that the literal sense of the divine Law is false and disclosing its allegorical sense would precisely undermine the intention of the prophet who concealed the allegorical sense because of the "difference in human

[18]Note that the pagination of the Arabic and the English are the same in the edition I used.

[19]For this argument, see in particular *Faṣl al-maqâl*, 8; 19; 24–25. Cf. *Kitâb al-kashf*, Ar. 132–135; Eng. 16–19.

nature." Averroes explains this by drawing an analogy between the role of the medical doctor and the role of the lawgiver in which he opposes the lawgiver to a person who intends to disclose the allegorical content of the divine Law:

> Here is a parable of these people's intention as contrasted to the intention of the Lawgiver [*al-shârі*']: Someone intends [to go] to a skilled physician who intends to preserve the health of all of the people and to remove sickness from them by setting down for them prescriptions to which there is common assent [*mushtarakat al-taṣdîq*] about the obligation of practicing the things that preserve their health and remove their sickness, as well as of avoiding the contrary things. He is not able to make them all become physicians, because the physician is the one who knows by demonstrative methods [*bi-al-ṭuruq al-burhânîyya*] the things that preserve health and remove sickness. Then this one [the allegorical exegete] goes out to the people and says to them: "These methods this physician has set down for you [. . .] have interpretations." Yet they do not understand [these interpretations] and thus come to no assent as to what to do because of them. (*Faṣl al-maqâl*, 27–28)[20]

To the "health" in the parable corresponds the perfection and happiness to which the prophet and lawgiver intends to lead all human beings to the extent they can attain it. To the "prescriptions" corresponds the divine Law. What Averroes means is that if the beliefs based on the literal sense of the divine Law are taken away from non-philosophers who do not understand the allegorical sense, because they lack the required intellectual abilities for understanding it, then these non-philosophers will fall into nihilism. For they will not follow the guidance of the lawgiver on account of its literal sense which has lost its authority for them, nor will they follow it on account of the allegorical sense, because they do not understand it. They loose, for instance, their belief in God as a king who rewards the obedient and punishes the disobedient. At the same time, they are unable to understand the notion of a first cause and how it relates to a virtuous life. Hence they loose both their belief in God and their belief in the value of a virtuous life. Again and again Averroes stresses that the allegorical sense of the divine Law is not to be made public. His sharp criticism of Muslim theologians who "strayed and led astray" is motivated above all by the fact that they "revealed their allegorical interpretation to the multitude [*saraḥû bi-ta'wîlihim li-l-jumhûr*]," i.e., did not respect the divisions due to the "difference in human nature."[21] The theologian must never go beyond the literal sense when he addresses non-philosophers. Like philosophy, the allegorical sense of scripture must remain concealed. As a consequence, philosophical doctrines may only be recorded in books that employ scientific demonstrations. For these, according to Averroes, are protected by their difficulty: books which "use demonstrations are accessible only to those who understand demonstrations" (*Faṣl al-maqâl*, 21). This, of course, is as true for Spinoza's *Ethics,* 450 years later, as it is for Averroes' commentaries on Aristotle.

---

[20]For the metaphor of the physician, see also *Kitâb al-kashf*, Ar.181; Eng. 67.

[21]See *Faṣl al-maqâl*, 29–32; to have shown that allegorical interpretation is strictly reserved to philosophers is, according to *Kitâb al-kashf*, Ar. 132–133; Eng. 16–17, one of the main results of the *Faṣl al-maqâl*.

It should have become clear that Averroes' version of dogmatism is not exposed to the three objections which I outlined above. Firstly, the contradictions between theology and philosophy that Spinoza stresses in the third letter to Blyenbergh simply follow from the fact that the arguments of theology are based on the literal sense of scripture. This implies by no means that for Spinoza the allegorical sense of scripture does not agree with the doctrines demonstrated in philosophy. As we saw earlier, he expressly states their agreement, among other places in *Cogitata Metaphysica* 2.8 and in the first letter to Blyenbergh. Theology, according to Spinoza, "with good reason represented God as a perfect man," who "is affected by anger through the deeds of the impious and delights in those of the pious." For theology's purpose is not to determine God's existence and essence philosophically, but to convey through rhetorical-poetical means an idea of God to non-philosophers and to guide them to virtuous action. Also, the second objection does not hold. It is clear now why an Averroist would not attempt to resolve contradictions between philosophy and theology by composing an allegorical commentary on problematic passages in scripture as Maimonides did, for instance, with Biblical anthropomorphisms. Finally, Averroists would also not institute an exegetical tyranny of philosophers. On the contrary: The philosopher is prohibited from intervening in the beliefs of popular religion even if they are philosophically as untenable as the anthropomorphic representation of God.[22] Averroes recognizes, of course, a set of fundamental religious principles to which all members of the religious community must subscribe. They include, for example, God's existence and unity. But these exist in Spinoza's *religio catholica* as well. He clearly does not extend freedom of opinion and interpretation to the principles of the *religio catholica*.[23]

On one important issue, however, Spinoza and Averroes differ. As we saw in the passage from *Cogitata Metaphysica* 2.8, Spinoza takes it for granted that scripture must "teach the same things" that we "grasp in the most certain way through natural reason." Unlike Averroes, however, Spinoza does not infer from this that philosophers are obligated to find an allegorical interpretation for every apparent contradiction between philosophy and scripture. It seems that for Spinoza it is sufficient to assume that in principle philosophy and scripture agree. I will come back to this issue below.

Deviations between Spinoza's position and the position of the *Faṣl al-maqâl* in part are simply due to the fact that Spinoza did not read Averroes' treatise. For one thing, it was not part of the Latin reception of Averroes. It is precisely because Averroes' philosophical-theological works were not known to the Latin West that he came to be represented as a philosophical heretic and denier of religion.[24] One only needs to read the article on Averroes in Pierre Bayle's *Dictionnaire historique et critique* to see that this distorted view of Averroes remained alive in the Early

---

[22]Cf. Stroumsa (2005), 20.

[23]Cf. *TTP* 14.

[24]Cf. Ivry (1988).

Modern period.[25] What I tried to characterize as Spinoza's Averroism has nothing in common with this tradition. It is, moreover, highly unlikely that Spinoza read the medieval Hebrew translation of the *Faṣl al-maqâl*.[26] No reference to it is found in Spinoza, nor is there any evidence that this translation was known in Jewish intellectual circles in the seventeenth century. We do know, on the other hand, that Spinoza owned a copy of the treatise *Behinat ha-dat* (Examination of Religion) by the Renaissance Averroist Elijah Delmedigo who made substantial use of the *Faṣl al-maqâl* to clarify the relationship between the Mosaic Law and philosophy.[27] As I suggested at the beginning, all parallels between Spinoza and Averroes that I pointed out can be explained on the assumption that Spinoza read Delmedigo's treatise. This assumption gains plausibility because other writings in the same volume containing Delmedigo's treatise left traces in Spinoza's work. It gains additional plausibility because the contradiction between philosophy and theology discussed in one of the three passages in Spinoza's early writings mentioned above corresponds precisely to the only example for such contradictions given by Delmedigo: the contradiction concerning the understanding of angels.[28] Finally, I will show below that where Delmedigo deviates from the orthodox Averroist position, Spinoza is clearly closer to Delmedigo than to Averroes.

In a paper published in 1922 Leon Roth documented the traces in Spinoza's work left by the volume containing Delmedigo's treatise. In the same paper he also drew attention to the importance of Delmedigo for understanding Spinoza. Roth's suggestion has not been pursued further by Spinoza scholars. In my view he not only misunderstood Delmedigo, but also misrepresented his influence on Spinoza:

> It is perhaps hardly necessary to point out how closely this [i.e., Delmedigo's position] is reproduced in the *Tractatus Theologico-Politicus*. The professed aim of the *Tractatus* is to refute the view of Maimonides that philosophy and theology are identical, and the crucial chapter to which all the earlier chapters are preliminary [i.e., chapter 15] sums up the discussion in the very words of the *Examination of Religion*. [...] The definite sundering of the spheres of theology and philosophy to the establishment of which [...] the *Tractatus Theologico-Politicus* is specifically devoted, is one of the landmarks in the history of political freedom as well as of intellectual development. [...] We now see that the very phraseology of its main thesis is to be found in the obscure Hebrew essay of R. Elijah. (Roth 1922, 58.)

A close reading of the *Behinat ha-dat* does not confirm Roth's thesis. As I understand Delmedigo he assumed, like Averroes, that religion and doctrines demonstrated in philosophy cannot contradict each other. This interpretation is, however, controversial among Delmedigo scholars, and I will briefly discuss the matter below. It is, by contrast, uncontroversial, that if Delmedigo allows for contradictions between religion and philosophy, the former always overrides the latter.

---

[25] See *Dictionnaire*, 384–391.

[26] For the Hebrew translation, see N. Golb (1956–57).

[27] See already Hübsch (1882–83). Cf. Ivry (1983) and Motzkin (1987). For scholars who claim that Delmedigo is closer to Latin Averroists than to Averroes, see below, n. 34.

[28] See *Cogitata Metaphysica* 2.12 and *Behinat ha-dat*, 93.

Both positions are incompatible with Spinoza's stance in the *TTP*. Let me now briefly examine how Delmedigo appropriates the fundamental assumptions of the *Faṣl al-maqâl*:

> And we say that adherents of religion who are correct in their views do not doubt that the purpose of the Law of Moses is to guide us in human affairs and in good deeds, as well as in true opinions insofar as this is possible for the entire people and according to the nature of the select few [*ha-yehidim*] with respect to what is their exclusive domain. Hence the Law of Moses and the prophets set down certain fundamental principles by way of tradition and by way of rhetorical and dialectical explanations in accordance with the method of assent [*mishpat ha-'immut*] that is characteristic of the multitude, and it [the Law of Moses] stirred the select few to investigate according to the method of assent characteristic of them concerning these issues [i.e., the demonstrative method]. [...] And the following becomes clear [. . .]: that the Law of Moses aims at the perfection of every adherent of religion insofar as possible to him. And since demonstrative science is impossible for the multitude as a whole, while it is possible for the select few—for this reason the Law of Moses requires both these things [i.e., assent on the basis of rhetorical and dialectical arguments and assent on the basis of demonstrative arguments]. (*Behinat ha-dat*, 76)

As in Averroes, the methods used by the Law of Moses for the guidance of non-philosophers—e.g. rhetorical and dialectical arguments—lead to contradictions with the teachings of philosophy. Delmedigo stresses from the outset that methods vary significantly from one discipline to another. The same Biblical text, for example, will be studied in different ways by a Talmudist whose goal is to arrive at a legal decision, by a grammarian whose goal is to provide evidence for a grammatical rule, and by an exegete whose goal is to clarify the text's meaning. The inference Delmedigo wants the reader to draw is clear: a prophet whose goal is to maximize the perfection of the religious community will speak differently about things like God, angels, or providence than a philosopher whose goal is to establish what is true and false.[29] While the prophet's methods are poetical, rhetorical, and dialectical, the philosopher uses scientific demonstrations. These goal-dependent differences in method can, but must not, lead to contradictions.[30] There is, for instance, no contradiction between prophetic and philosophical statements concerning God's existence and unity.[31] For the prophet, however, the scope of true opinions which he can communicate and the quality of the proofs on which he can ground them are limited by his overall goal: to establish the moral, political, and intellectual conditions for perfection in a community made up of philosophers and non-philosophers. If the goal-dependent differences in method give rise to contradictions, Delmedigo argues, one way of resolving them is through allegorical exegesis. There are cases, he argues, in which "a thing has an interpretation reserved to the select few"

---

[29]On the goal of the Mosaic Law, see *Behinat ha-dat*, 75–76; on the difference between the Mosaic Law and philosophy with respect to method, see in particular 92–94.

[30]Strictly speaking, these are different methods belonging to the same discipline, i.e., logic. On the inclusion of the *Rhetoric* and *Poetics* into Aristotle's *Organon* and its philosophical implications, see Black (1990). Delmedigo (*Behinat ha-dat*, 75) briefly refers to the different methods of "logic" (*ha-limmud ha-kolel*).

[31]See *Behinat ha-dat*, 76–78.

(*Behinat ha-dat*, 77). One such case concerns angels: for philosophers they are entities "assumed to be separate from any body and corporeal attribute." In other words: they are the incorporeal intelligences of the supralunar world as conceived by medieval Aristotelians. In the Bible, by contrast, angels are described as entities "apprehended through sense-perception as we apprehend bodies" (*Behinat ha-dat*, 93). This, of course, is a concession to non-philosophers who are not familiar with the physical and metaphysical proofs for the existence and the attributes of incorporeal intelligences. If the prophet arrives at the conclusion that in order to attain his overall goal it is required to convey a notion of angels to non-philosophers, he must present them within a conceptual framework that his audience can understand. Like Averroes, Delmedigo harshly criticizes the disclosing of such allegorical interpretations in public:

> Many of those who philosophize among the people of our nation have in my opinion strayed from the method of the Law of Moses and its intention. And this is because they sought to change all the literal meanings of the verses [*peshate ha-pesuqim*] which are [found] in most of the branches and stories of the Law of Moses, as if they wished to make the words of the Mosaic Law more beautiful and to ground them on the meanings [inferred by] scientific syllogism [*ha-heqqesh ha-sikhly*]. And they did not succeed in either this or that [...], and I think that this should not be done at all. [...] My method, therefore, is very different from the method of many who philosophize in our nation. They changed the goal both of the Mosaic Law and of philosophy and mixed the two [kinds] of investigation—the theological and the speculative [*ha-torani ve-ha-'iyyuni*]—together, as well as the universal and the specific method [*ha-derekh ha-kolel ve-ha-miyyuhad*]. And they are like intermediaries between the theologians [*ha-medabberim*] among the religious people and the philosophers. (*Behinat ha-dat*, 93–94.)

Delmedigo explicitly mentions Maimonides in this context as someone who "walked on the way" that he has criticized (*Behinat ha-dat*, 84). As I mentioned above, Delmedigo attaches great importance to the fact that prophecy and philosophy each have their own goal and as a consequence use different methods to attain it. While the method of the philosopher is "universal"—establishing what is true and false on the basis of scientific demonstrations which are valid always and everywhere—the method of the prophet is "specific"—establishing the moral, political, and intellectual conditions for human perfection in a religious community shaped by a particular set of geographic and cultural circumstances. If according to the prophet circumstances require presenting angels to non-philosophers in corporeal terms, the purpose of doing so would be undermined if a philosopher disclosed in public that, correctly understood, this account refers to incorporeal intelligences. The philosopher would be disregarding the political considerations that led to the allegory in the first place.[32] Like Averroes in the analogy between the lawgiver and

[32] According to Delmedigo, the disclosure of the allegorical interpretation of angels led to conflict and strife between philosophers and kabbalists in the Jewish community (see Delmedigo, *Behinat ha-dat*, 93–94). His account of the conflict is clearly modelled on Averroes' description of the emergence of factions in Islam as a consequence of the disclosure of allegorical interpretations. See *Faṣl al-maqâl*, 29–32.

the doctor that we saw above, Delmedigo stresses the danger inherent in disclosing the allegorical content to non-philosophers:

> When we tell these deep things [*'eleh ha-'amuqot*] as they truly are to the multitude, we do not benefit them, for they do not understand them, but we cause them great damage. (*Behinat ha-dat,* 96)[33]

It would, therefore, be clearly a mistake to publicly interpret verses in the Law of Moses which conflict with doctrines demonstrated in philosophy. To quote once more Spinoza's third letter to Blyenbergh: theology presented God anthropomorphically "with good reason." In order not to undermine this "good reason" the scientific examination of God must be confined to philosophical treatises. This does not, however, mean that contradictions cannot *in principle* be resolved through allegorical interpretation. At no point does Delmedigo question the truth of the Mosaic Law.

Until now I have portrayed Delmedigo as an orthodox Averroist. This portrait, however, requires two modifications. Firstly, there is the view, persistently reiterated in the scholarly literature, that Delmedigo, in contrast to the historical Averroes, was not committed to the "identity of religious and scientific truth," but "obviously" adopted the theory of double truth that allegedly was set forth by Christian Averroists.[34] I have shown elsewhere why this interpretation of Delmedigo is implausible.[35] For my present purpose a brief summary of my argument must suffice.[36] The case Delmedigo considers is the conflict between two positions of which neither can be conclusively demonstrated. In this case the philosopher will choose the side which is *most likely* in light of the available evidence. Since the available evidence may change as a consequence of scientific progress, the position that was less likely at one point may become more likely at another. If such a conflict occurs between a position advocated in philosophy and a fundamental principle of the Mosaic Law, it cannot be resolved on scientific grounds—assuming, of course, that the philosophical position was established by sound scientific methods. It can also not be resolved on exegetical grounds, because fundamental principles are not open to interpretation: they are necessary conditions for achieving the purpose of the Law of Moses. Since in his scientific investigations a Jewish philosopher must rely on sound scientific methods, he is led to adopt the philosophical position. On the dogmatic assumption that the Law of Moses is true, he will at the same time remain convinced that once *all* evidence becomes available, the position of the Law of Moses will be vindicated. Philosophical and religious commitments thus can be at variance *temporarily* on account of the contingent state of scientific knowledge. *Absolutely* speaking, however, they must be in agreement. It is true that Averroes did not consider such a case. Delmedigo's model here is most

[33]Note that this passage comes in the context of Delmedigo's discussion of rabbinic *aggadot*.

[34]This interpretation was first proposed by Julius Guttmann in a critical response to Hübsch (see above, n. 27). The quotation is from Guttmann (1927), 197–198. It was reiterated by Geffen (1973–74) and Ross (1984), 48–54; Ross's assessment is the most differentiated to date.

[35]See Fraenkel (forthcoming).

[36]What follows is my understanding of Delmedigo's position set forth in *Behinat ha-dat*, 77–85.

likely Maimonides' account of the conflict between the Mosaic Law and Arabic Aristotelians on the question whether the world is created or eternal.[37] But none of this supports the claim that Delmedigo abandoned the fundamental assumption of dogmatism concerning the "identity of religious and scientific truth."

More interesting for my present purpose is the second point on which Delmedigo deviates from the orthodox Averroist position. In the *Faṣl al maqâl*, Averroes not only assumes that every contradiction between the divine Law and philosophy can *in principle* be resolved through allegorical interpretation, but that the philosopher *is obligated* to resolve contradictions in this manner.[38] But what is the benefit derived from doing so given the strict prohibition to disclose allegorical interpretations? Why is it not sufficient if the philosopher is in principle committed to the agreement between the divine Law and philosophy? While Delmedigo allows for allegorically resolving contradictions as long as they are not contradictions of the type just outlined, he is clearly not enthusiastic about doing so. Carrying out such interpretations is, as it were, useless and, in addition, dangerous if the interpretations are disclosed in public. The best way of studying the propositions of the Mosaic Law is in light of the Law's own peculiar methods and purpose. The aim then would be to understand how these propositions contribute to maximizing the perfection of the religious community. Instead of working out how the anthropomorphic representation of angels, for instance, allegorically refers to incorporeal intelligences, the question becomes which political considerations motivated Moses to represent angels in such a way. Seeking the allegorical content of the Mosaic Law would mean to study it in view of establishing the truth which is the goal of philosophy. This would be as pointless as making poetical, rhetorical, or dialectical arguments in a philosophical treatise in view to communicating its content to non-philosophers, which is the goal of prophecy. Concerning miracles, for instance, Delmedigo explicitly questions the purpose of changing the literal meaning of the Mosaic Law, since both philosophers and non-philosophers accept them, even though they understand them in different ways. It is thus not surprising that he implicitly casts doubt on the philosopher's obligation to provide allegorical explanations. The philosopher should "perhaps" (*'ulay*) interpret such passages in scripture that, taken literally, contradict doctrines demonstrated in philosophy (*Behinat ha-dat*, 93). Delmedigo thus puts more stress than Averroes on the methodological autonomy of philosophical and prophetic discourse. But this does not mean that he is less committed to the fundamental assumption of dogmatism concerning the agreement of philosophy and religion.

As I already suggested, Delmedigo's deviation from the orthodox Averroist position supports my claim that he is the source of what I described as Spinoza's Averroism. For already in his early writings Spinoza goes one step further than

---

[37] See in particular *Guide* 2.13–25. For the concept of scientific progress, see in particular 2.19 and 2.24. For considerations of probability, see 2.23. Note that Delmedigo is critical of Maimonides' attempt to settle the matter through scientific arguments.

[38] See *Faṣl al-maqâl*, 9–10 and 19–20.

Delmedigo: he drops the obligation to provide allegorical interpretations altogether. Recall once again the passage from *Cogitata Metaphysica* 2.8: "here we only inquire into the things which we can grasp in the most certain way through natural reason; and it is sufficient that we demonstrate these clearly in order to know that scripture must also teach the same things." Thus in order to ground the authority of scripture dogmatically Spinoza considers it sufficient to assume that its allegorical content can in principle not contradict what is clearly demonstrated by natural reason. There is no need to actually seek for the allegorical content. Finally, the position advocated in the *TTP* in one sense can be understood as a further radicalization of the methodological autonomy of philosophy and religion assumed in the Averroistic tradition. In another sense, however, Spinoza in the *TTP* breaks with the fundamental premise that underlies the dogmatism not only of al-Fârâbî, Maimonides, Averroes, and Delmedigo, but also of his own early writings: that "the truth does not contradict the truth."

References

Black, Deborah. 1990. *Logic and Aristotle's "Rhetoric" and "Poetics" in medieval Arabic philosophy*. Leiden: E. J. Brill.

Fraenkel, Carlos. 2008a. Could Spinoza have presented the *Ethics* as the true content of the Bible? *Oxford Studies in Early Modern Philosophy* 4, eds. Daniel Garber and Steven Nadler, 1–50.

Fraenkel, Carlos. 2008b. Philosophy and exegesis in Al-Fârâbî, Averroes, and Maimonides. *Laval Théologique et Philosophique* 64: 105–25.

Fraenkel, Carlos. 2010. Legislating truth: Maimonides, the Almohads, and the 13th century Jewish Englishtenment. *In Studies in the history of culture and science presented to Gad Freudenthal*, eds. Resianne Fontai et al., 209–23 Leiden: Brill.

Fraenkel, Carlos. Forthcoming. Reconsidering the case of Elijah Delmedigo's Averroism and its impact on Spinoza. In *Renaissance Averroism and its aftermath: Arabic philosophy in early modern Europe*, eds. Anna Akasoy and Guido Guiglioni.

Geffen, David. M. 1973–74. Insights into the life and thought of Elijah Del Medigo based on his published and unpublished works. *Proceedings of the American Academy for Jewish Research* 61–2: 69–86.

Goetschel, Willi. 2004. *Spinoza's modernity*. Madison, WI: University of Wisconsin Press.

Guttmann, Julius. 1927. Elias del Medigos Verhältnis zu Averroes in seinem Bechinat ha-Dat. In *Jewish Studies in Memory of Israel Abrahams*, ed. A. Kohut, 192–208. New York: Press of the Jewish Institute of Religion.

Hübsch, Adolph. 1882–3. Elia Delmedigos Bechinat ha-Dath und Ibn Roshd's Facl al-maqal. *Monatsschrift für Geschichte und Wissenschaft des Judentums* 31: 552–63, 32: 28–48.

Israel, Jonathan. 2001. *Radical enlightenment*. Oxford: Oxford University Press.

Israel, Jonathan. 2006. *Enlightenment contested*. Oxford: Oxford University Press.

Ivry, Alfred. 1983. Remnants of Jewish Averroism in the Renaissance. In *Jewish thought in the sixteenth century*, ed. Bernard. D. Cooperman, 243–265. Cambridge: Harvard University Center for Jewish Studies.

Ivry, Alfred. 1988. Averroes and the West: The first encounter/non-encounter. In *A straight path: Studies in medieval philosophy and culture*, ed. Ruth Link-Salinger, 142–158.Washington, DC: Catholic University of America Press.

Motzkin, A. L. 1987. Elija del Medigo, Averroes and Averroism. *Italia* 6: 7–20.

Pines, Shlomo. 1963. Translator's Introduction. In *The Guide of the Perplexed*, by Maimonides, trans. Shlomo Pines, lvii–cxxxiv. Chicago: University of Chicago Press.

Ross, J. J. 1984. Introduction [Hebrew]. In Elijah Delmedigo, *Sefer behinat ha-dat. Critican edition with introduction and commentary by J. J. Ross*. Tel Aviv: Chaim Rosenberg School of Jewish Studies.

Roth, Leon. 1922. The *Abscondita Sapientiae* of Joseph del Medigo. *Chronicon Spinozanum* 2: 54–66.

Stroumsa, Sarah. 2005. Philosophes Almohades? Averroes, Maimonide, et l'idéologie almohade. In *Los Almohades: Problemas y perspectivas*, eds. P. Cressier, M. Fierro and L. Molina, 1137–1162. Vol. 2. Madrid: CSIC.

Bibliography of Primary Sources

Al-Fârâbî. 2005. *Kitâb al-hurûf* [The Book of Letters]. Edited by M. Mahdi. Beirut: Dâr al-mashriq, 1969, Eng. trans. of Book 2 in *Medieval Islamic philosophical writings*. ed. Muhammad Khalidi. Cambridge: Cambridge University Press.

Al-Fârâbî. 2001. *Kitâb al-milla* [The Book of Religion]. Edited by M. Mahdi. Beirut, 1968. Eng. trans. by C. Butterworth in *The Political Writings: Selected Aphorisms and Other Texts*. Ithaca: Cornell University Press.

Al-Fârâbî. 1962. *Taḥsîl al-sa'âda* [The Attainment of Happiness]. Edited by J. Al-Yasin, Beirut, 1981. Eng. trans. by M. Mahdi in *Al-Farabi's Philosophy of Plato and Aristotle*. Ithaca: Cornell University Press.

Averroes. 1974. *Commentary on Plato's Republic*. Translated into Hebrew by S. B. Judah of Marseilles. Edited with Eng. trans. by E. I. J. Rosenthal. Cambridge: Cambridge University Press, 1969. Also translated by R. Lerner as *Averroes on Plato's Republic*. Ithaca NY: Cornell University Press.

Averroes. 2001. *Faṣl al-maqâl* [Decisive Treatise]. Ed. George Hourani, trans. Charles Butterworth, with corrections by Muhsin Mahdi. Provo: Brigham Young University Press.

Averroes. 1956 and 1957. *Faṣl al-maqâl* [Decisive Treatise]. Edited with Hebrew translation, N. Golb, as "The Hebrew Translation of Averroes' Faṣl Al-Maqâl" in *Proceedings of the American Academy for Jewish Research* 25: 91–113, 26: 41–64.

Averroes. 2001. *Kitâb al-kashf* [The Book of Exposition]. Edited by M. Qasim. Cairo, 1964. Eng. trans. by I. Najjar in *Faith and Reason in Islam*. Oxford: Oneworld.

Bayle, Pierre. 1697 and 1730. *Dictionnaire historique et critique*. 1. Edition, 1697. 4. Edition, Amsterdam: Brunel et al., 1730.

Delmedigo, Elijah. 1984 *Sefer behinat ha-dat* [The Examination of Religion]. Edited by J. J. Ross. Tel Aviv: The Chaim Rosenberg School of Jewish Studies.

Maimonides. *Dalâlat al-hâ'irîn* [Guide of the Perplexed]. Edited by S. Munk and Y. Yoel. Jerusalem, 1931. Translated into Hebrew by S. ibn Tibbon as *Moreh ha-Nevukhim*, edited by Y. Even-Shmuel. Jerusalem: Mosad ha-Rav Kook, 1987. Eng. trans. S. Pines, Chicago: University of Chicago Press, 1963.

Spinoza, Benedict. *Opera*. Edited by Carl Gebhardt. 4 vols. Heidelberg: Carl Winters Universitätsbuchhandlung, 1925.

Spinoza, Benedict. *Tractatus-Theologico Politicus*. Edited by Fokke Akkerman with French translation by Jacqueline Lagrée and Pierre-François Moreau in *Oeuvres*. Vol. 3. Paris: Presses Universitaires de France, 1999.

Chapter 4
Music, Mechanics and "Mixed Mathematics"

Alison Laywine

4.1

It might seem odd to contribute a paper on music theory to this volume. But it is less so than the reader might think. Music theory was much more closely related to science and philosophy in the early modern period than it is today. This is not simply because intellectual life was less specialized back then; it reflected a wide-spread, traditional understanding going back to Greek Antiquity that music theory was itself a science. It was treated as a branch of "mixed mathematics," i.e., as a science that mixes mathematics with the relevant phenomena. Thus it historically had the same status as its sister science astronomy. The task of astronomy was to seek the rational order in the phenomena of seeing, i.e., in the observed motions of the heavens; the task of music theory was to seek the rational order in the phenomena of hearing, i.e., in the phenomena of consonance and dissonance—the blending or clashing of notes of different pitch.[1] The preface to the 1569 edition of Niccolò Tartaglia's Italian presentation of Euclid expresses this conception of music theory very clearly:

> We know that all the other sciences, arts, and disciplines need mathematics; not only the liberal arts, but all the mechanical arts as well. ... And it is also certain that these mathematical

A. Laywine (✉)
McGill University, Montreal, QC, Canada
e-mail: a.laywine@mcgill.ca

I presented earlier versions of this paper at a conference in May 2004 at the University of Toronto and then in October 2006 at the University of New Hampshire in Durham, New Hampshire. I am indebted to the stimulating discussion that followed my presentation in both places. I am especially indebted to feedback from Calvin Normore at the Toronto event. This paper has also benefited from stimulating discussion with Stephen Menn. I also extend very special thanks to Peter Schubert, who shared with me his expertise both as a scholar of Renaissance music theory and as a choral conductor of Renaissance polyphony.

[1] The ancient texts that express this idea and were taken by later people to define the place of music theory in the mathematical curriculum are Fragment One attributed to Archytas (for the Greek text, see Düring (1932, 56.5–57.27) and Plato, *Republic*, Book VII, 530d (for the Greek text, see Burnet (1992). The Archytas, as translated by Andrew Barker reads as follows: "Those who are concerned with the sciences seem to me to be men of excellent discernment, and it is not strange that they

C. Fraenkel et al. (eds.), *The Rationalists: Between Tradition and Innovation*,
The New Synthese Historical Library 65, DOI 10.1007/978-90-481-9385-1_4,

> sciences or disciplines are the nurses and mothers of the musical sciences, since it is with numbers and their properties, ratios and proportions that we know the octave, or double ratio [sc. 2:1], to be made up of the ratios 4:3 and 3:2; and it is similarly that we know the former [sc. the musical interval corresponding to the ratio 4:3, i.e., the fourth] to be composed of two tones [sc. (9:8)(9:8)] and a minor semitone [sc. 256:243], while the latter [sc. the musical interval corresponding to the ratio 3:2, i.e., the fifth] is composed of three tones and a minor semitone. And thus the octave (or double) is composed of five tones and two minor semitones; that is, a comma [sc. 531441:524288!] less than six tones; and likewise we know a tone to be more than eight commas and less than nine. Also, by virtue of those [mathematical] disciplines, we know it to be impossible to divide the tone, or any other superparticular ratio [sc. any ratio of the form n+1:n], into two equal [sc. rational] parts [sc. in geometric proportion], which our Euclid demonstrates in the eighth proposition of Book VIII.[2]

Now Tartaglia was an engineer, a gunnery expert, and a private tutor in mathematics, but he was not a professional musician.[3] So one might expect his account of the "musical sciences" in the passage just quoted to be idiosyncratic. But it is not. This can be seen quite clearly if we compare it with that of Gioseffo Zarlino (1517–1600)—choirmaster at Saint Mark's in Venice and, among professional musicians, perhaps the single most influential music theorist of the time. In his great treatise of counterpoint, *Le istitutioni harmoniche* of 1558, Zarlino says that theoretical or speculative music

> cannot be said simply to be a mathematical science, nor can it be said simply to be a natural science, but rather it should be said to be partly one and partly the other and therefore a mediator between them. And since it is from natural science that it gets the account of the matter of consonance (the matter of consonance being sounds and voices) and since it is from mathematical science that it gets the account of the form of consonance (the form of consonance being its proportion [sc. 2:1 for the octave, 3:2 for the fifth, 5:4 for the major third, etc.]) therefore, since it is right to name things after their nobler side, it is more reasonable to say that music is a mathematical science than that it is a natural science, if only to the extent that the form is nobler than the matter. (Zarlino 1573, Pt. 1, Chap. 20, 38)

Precisely because music theory was treated as a branch of mixed mathematics, it drew the interest of people from lots of different disciplines and drew them into debate with one another. Naturally it attracted the interest of professional musicians who wanted to understand the scientific foundations of the rules of counterpoint: the students and critics of Zarlino. But it also attracted the interest of people broadly

---

conceive particular things correctly. For since they exercised good discrimination about the nature of the wholes, they were likely also to get a good view of the way things really are taken part by part. They have handed down to us a clear understanding of the speed of the heavenly bodies and their risings and settings, of geometry, of numbers, and not least of music. For these sciences seem to be sisters, since their concern is with the two primary forms of what is, which are sisters themselves" (Barker 1989, 39–40). The Plato passage, as translated by Barker, reads as follows: "'It appears,' I [Socrates] said, 'that just as the eyes are fixed on astronomy, so the ears are fixed on harmonic motion, and that these two sciences are one another's sisters, as the Pythagoreans say and we agree, Glaucon" (1989, 55).

[2]Quoted in translation by Drake (1970, 47).

[3]For a brief account of Tartaglia, see Drake (1970, 26 ff.). Tartaglia's dates seem to be 1500 to 1557.

concerned with the mathematization of physical phenomena. For example, Galileo devoted to it the final pages of Day One of the *Discourses on Two New Sciences* (1636). Isaac Beeckman had a long-standing interest in the subject and introduced Descartes to it some time after they met in 1618. It was under Beeckman's influence that Descartes wrote his first book—a work on music theory called the *Compendium musicae*. The *Compendium musicae* is probably Descartes' *least* important book. But the *most* important book that *Marin Mersenne* ever wrote is on music theory: it is a vast, rambling production in three volumes called *L'Harmonie universelle*. Mersenne believed that music theory had some very special significance. That's already obvious from the sheer size of his book, but also from programmatic remarks he makes such as the following:

> Sounds can shed more light on philosophy than any other quality. This is why the science of music must not be neglected, even if singing and instrumental music were entirely abolished and forbidden. For they [sc. singing and instrumental music] are not the chief end of music .... In effect, if the knowledge of sounds and their proportion can open the door to the propositions concerning the objects of the eye, smell, and taste, every worthy man would prefer this knowledge to all the singing and all the instrumental music that can be performed according to the rules of art. (Mersenne 1666, Prop. 33, corollaire 3, 88)

As all of this goes to show, not only was music theory taken to be a science in the sixteenth and seventeenth centuries, it was a science with some claim on the attention of a very wide and learned audience.

Debates about music in the early modern period are philosophically interesting for a number of reasons—but particularly in the way they raise questions in the philosophy of science. Parties involved in the debates found they had to think long and hard about what should and should not count as scientific knowledge, how such knowledge can be acquired in the fields where theory and observation cross-pollinate, what should be admitted as evidence, how disputes may be settled, whether there is one or several criteria of truth for settling disputes in musical science and, if so, how it or they should be understood. These questions were very pressing because the debates were often very heated. The debates were heated for the reason that they always are in music—as they are even today in some quarters: people passionately care about what is right and wrong in music almost as much as they care about what is right and wrong in ethics and politics.

Now the underlying assumption of the music theorists of the early modern period is alien to us—even to those of us who haven't thought about scales and arpeggios since long forgotten childhood music lessons. We no longer think that music theory is a science at all, even if we grant that it is rigorous and precise in its way—perhaps for the following reason. It probably seems reasonable to most of us today to assume that the object of music theory is not some part of nature subject to natural laws, but rather a cultural artifact determined by musical taste that can change over time and from culture to culture. There can't be an exact science of anything determined by arbitrary choice. If that's where we come down, we will find it hard to understand sympathetically where a lot of early modern music theory is coming from. But we can use some of the *other* assumptions we likely have about music theory today, to try to develop an initial, hypothetical idea of what might make good sense to us, if we *did* believe that music theory is a science. So here is another assumption I think many of us would be willing to make—again, even if we long ago abandoned our childhood music teachers: music theory better have *something* to say about the music people make. If you share this intuition with me, you will naturally expect to discover that the criterion of truth for music theory (for those who may have sought such a thing at any period in its history)[4] will be how the music sounds to the ear. If our theory defends principles that just sound bad sung or played, then something must be wrong with the theoretical assumptions. Given this very natural assumption about the relation between music theory and musical practice, we would expect to discover historically that the music theorists of the early modern period—who did treat their discipline as a science—must have thought the ultimate criterion of truth for settling disputes would be observed data. We would expect them to treat musical performance as a kind of laboratory experiment. If you tell me that the octave and a fourth is dissonant and I say otherwise on theoretical grounds (whatever they might be), we will argue until we are blue in the face—unless somebody comes along and plays an interval that you and I can both agree is an octave and a fourth. Then we will talk—more calmly, I hope—about whether we think it sounds good or not.

There has been on the table for a number of years now a scholarly account (better: a family of such accounts) of the music theoretical debates in the early modern period that tend to satisfy the expectations I just tried to raise—namely that music theory must have developed as an experimental science if it was treated as any kind of science at all. You can find this lesson drawn in essays by people like Claude Palisca (1961) Stillman Drake (1970) and Floris Cohen (1984). In the telling of their story, these people also satisfy an expectation that many of us will have of philosophical debates (broadly construed) in the early modern period—namely that these debates typically played themselves out in opposition to some long entrenched, intellectual authority. Just as we have long been accustomed to thinking of the many self-stylized "new philosophers" (Bacon, Hobbes, Descartes, or whoever) as struggling against the false authority of the "old philosophers" (Aristotle, the peripatetics, and medieval scholastics), so Palisca and friends introduce us to a new, fresh generation of musical thinkers (with some overlap of names from the list of new philosophers!) struggling against the forces of reaction in the person and writings of the aforementioned Zarlino. They will tell us the story, for example, of Vincenzo Galilei trying to undercut Zarlino's authority by putting his theories to empirical tests. "We just don't sing according to the principles you lay down," he will say to his former teacher, "and musical performance shows it."[5]

---

[4]I am borrowing the notion of a criterion of truth from Ptolemy. Thus, supposing music theory is a science after all, and granted that disputes can arise among us about what is true and what is false in this science, the criterion—whatevever it turns out to be at the end of the day—will be that authority to which we agree to appeal to resolve our dispute. In fact, it might be claimed that the debates in music theory of the sixteenth and seventeenth centuries I will be discussing in what follows can be understood as debates about what should count as the criterion of truth in music—taking this notion of criterion in Ptolemy's sense.

[5]Here especially see Drake's article "Vincenzio Galilei and Galileo" in his (1970).

Now this is what I propose to do. I would like to test the idea that I tried to make plausible, namely that music theory was an *experimental* science, if it was any kind of science at all. Since there already exists a scholarly account that plays up this idea, at least for the period in question, my strategy is to see how well it succeeds. The account offered to us by Palisca and friends ultimately turns on the insight that Zarlino was some kind of lightning rod in early music theory. There is no doubt that this insight is on target: any attempt to sort out this history and its larger philosophical significance has to come to terms with him. But I am going to challenge the picture of Zarlino offered to us by Palisca and friends and reassess his significance for the development of music theory as an experimental science. This will help me correct what I take to be a misleading simplification in the Palisca story as a whole; and, at the end of my paper, I will use my correction of this story to offer a friendly amendment to Stillman Drake's speculations about the possible significance of music theory for the emergence of the new mechanics as a science in its own right. Let me begin by giving you a thumbnail sketch of Zarlino, what Palisca and friends are on to, and how they have got him wrong.

Gioseffo Zarlino was born outside of Venice in 1517. He was a student of Adrian Willaert who was choirmaster at Saint Mark's in Venice from 1527 to 1562. Zarlino was elected to this same position in 1565 after Willaert's immediate successor was called away to Parma. Willaert was one of the younger masters of Flemish choral polyphony: a music sung typically in four or more different parts stitched together according to the rules of modal counterpoint. It was in this tradition that Zarlino was schooled; it was this tradition that his writings sought to explain and defend theoretically. To get a sense of Zarlino, it is useful to contrast Flemish polyphony with the music of Greek Antiquity, as we know it from the theoretical texts that have come down to us. Ancient Greek music treated only the octave, fifth, and the fourth (and the intervals composed of these intervals and an octave) as concords, i.e., as intervals agreeable to the ear. Flemish polyphony, by contrast, rejected the fourth (that is, it rejected the fourth in the bass), but it embraced the major and minor thirds and sixths. These intervals would have been rejected by the Greeks as dissonant. Now the music theorists accepted as authorities by polyphonic composers in the generations before Zarlino (Boethius, Guido d'Arezzo, Gaffurio, Glarean) promoted a certain tuning system—i.e., a certain prescription for the size of the intervals acceptable in music and the rules by which they may be combined—that they took from the theoretical writings of Greek Antiquity. This system goes by different names. We will call it the Pythagorean system: it is attested, for example, in a fragment of Philolaus.[6] It takes the form of a diatonic scale formed by two equal semitones, expressed as the numerical ratio of 256:243 (corresponding to string lengths), and five equal whole tones, expressed numerically as the ratio 9:8. (You will notice that this is the scale described in the passage from Tartaglia.) Zarlino's

[6]This is 44 B6 in Diels and Kranz (1989).

merit, among other things, was to have recognized that this tuning system could not accommodate the polyphony of his day. I can briefly show you what he was thinking.

Take the example of the major third. This interval is equivalent to two whole tones. Two *Pythagorean* whole tones put together and expressed as a numerical ratio give us (9:8)(9:8) = 81:64. But the pure major third—the one embraced by Zarlino on behalf of the polyphony of his day—is equivalent to 5:4, an interval that, as you can see just from the numbers, must be somewhat smaller or flatter than the Pythagorean third: that's because the ratio 81:64 is bigger than 5:4. You can see that when we compare the two ratios in terms of 64. The ratio 5:4 then gets expressed as 80:64. So the Pythagorean third is correspondingly bigger, wider, or sharper than Zarlino's third—and it's noticeable—by the ratio of 81:80 (the "syntonic comma"). So Zarlino rejected the Pythagorean tuning system, because it would produce unpleasant thirds and sixths. But what to replace it with?

Zarlino read, or had access to reports of, the music treatise called the *Harmonics* by the Alexandrian astronomer Claudius Ptolemy of the second century AD. There he discovered an alternative diatonic tuning system that accommodates both the major and minor thirds and the major and minor sixths. This system is called the "syntonic" or tense diatonic. This system again takes the form of a scale consisting of two equal semitones, each expressed as the ratio of 16:15, three whole tones expressed as the ratio of 9:8, and two slightly flatter whole tones corresponding to the ratio of 10:9. It can seem really weird to us to have whole tones of different sizes in a scale, but that all by itself is not an argument against the system. The virtue of the system, from Zarlino's point of view, is that it can give us major thirds: (10:9)(9:8) = 5:4; minor thirds: (9:8)(16:15) = 6:5; major sixths: (5:4)(16:15)(9:8)(10:9) = 5:3; minor sixths: (6:5)(16:15)(9:8)(10:9) = 8:5. Of course, the system also gives us perfect fourths, fifths and the octave—(that was, among other things, what it was originally intended to do).[7]

So much for the thumbnail sketch. Now there is no question that Zarlino was the musical authority to reckon with in the early modern period, that he came under heavy fire from forward-thinking types with alternative tuning systems, and that he was regarded by these people as an obstruction: both to truth and new musical idioms. But there is something fundamentally mistaken in the way that Palisca and friends represent Zarlino.

The problem with Zarlino, they say, is that he was a "Pythagorean"—not in the sense of the tuning system we just looked at, but rather in what we might as well call the loose and popular sense. In other words, they mean simply that he had a fixation on numbers—not just any numbers either, but the "sonorous" numbers from one to six, the so-called "scenario." It's from these numbers that the ratios corresponding to the consonant intervals are formed: 5:3 for the major sixth, 6:5 for the minor third,

[7]Thus see Part Two, chapters 28 and 31 of the *Institutioni harmoniche* (Zarlino 1573, 114–118; 123–126).

5:4 for the major third, 4:3 for the fourth, 3:2 for the fifth and 2:1 for the octave.[8] Zarlino was struck by the fact that 6 is the first "perfect" number, i.e., the first of the natural numbers equal to the sum of its own factors: 1+2+3 = (1)(2)(3).[9] On the strength of this, say Palisca and friends, he ascribed mystical properties to the first six whole numbers. So wedded was he to these numbers, and to the musical ratios formed from them, that he preferred to stand by the numbers—even when they could not adequately account for musical practice (the most practical and playable tuning of keyboard instruments and fretted instruments like the lute seemed to require some kind of deviation from Ptolemy's syntonic diatonic—Zarlino himself knew this). Thus according to Palisca and friends, Zarlino was a theoretical tyrant with a penchant for numerology. His a priori theorizing was discredited by appeal to observation and experimentation with tuning on musical instruments and acoustical devices.

Here especially Vincenzo Galilei is the hero for Palisca and friends—but more especially for Stillman Drake. Irritated by Zarlino's theoretical dogmatism, Vincenzo tried to prove that there is nothing acoustically special in the numbers of the scenario at all. There is a famous legend most of us have heard, reported by Nicomachus, according to which Pythagoras discovered the concords when he heard a blacksmith hammering away on anvils of different sizes.[10] The legend also tells us that Pythagoras discovered the numerical ratios associated with the concords by experimenting with strings. He is supposed not just to have found these ratios for strings of equal thickness and tension, but different lengths (for these variables, 2:1 does indeed correspond to the octave, 3:2 to the fifth and 4:3 to the fourth); he is also supposed to have kept the length and thickness equal and to have varied the tension by suspending different weights to the strings. According to the legend, the numbers worked out either way to the familiar ratios formed from 1, 2, 3 and 4—just as Zarlino would have it. Vincenzo must have remembered this legend, and in his effort to discredit Zarlino he reasonably asked: what would happen if you actually performed an experiment with weights, as Nicomachus reports? He tried it out and discovered that the numbers are not what you would expect. The pitch ratios are not directly related to the ratios of the weights, but rather to the *square roots* of the weights. Not only do these numbers not come from the scenario, they are—with the exception of the square-root of four—irrationals! Thus does

---

[8]The minor sixth is a problem for him, since it corresponds to the ratio 8:5. Alas, eight does not belong to the senario. Zarlino gets around this, by treating the minor sixth as a composite interval: what you get by putting together a perfect fourth and a minor third: (4:3)(6:5) = 8:5. Thus the minor sixth has sonorous numbers in its ancestry, since three, four, five and six, the numbers assigned to its constitutive intervals, belong to the senario.

[9]Thus see, in particular, Part One, chapters 12 through 15 of the *Institutioni harmoniche*, pages 27–33 in the edition of 1573.

[10]See chapter 6 of the *Enchiridion*. An English translation can be found in Barker (1989, vol. 2). For the Greek text, see von Jan (1985 245, 248).

level-headed scientific common sense, with a talent for experimentation, triumph over Pythagorean superstition.[11]

This story is compelling in lots of ways; and, it points beyond music theory itself—partly because this Vincenzo Galilei was Galileo's father. Thus it really is tempting to think, as Stillman Drake has argued, picking up Palisca's cue, that the precedent for Galileo's own experimental testing of the famous law of falling bodies is probably to be found in Vincenzo's polemics against Zarlino: if Drake is right, there doesn't seem to be any earlier precedent.[12]

But having said all that on behalf of Palisca and friends, the picture of Zarlino as an arch-Pythagorean in the loose and popular sense overlooks something potentially quite interesting. I'm going to argue that Zarlino was not just committed to Ptolemy's tuning system; he was also *at least verbally* committed to Ptolemy's research program in music theory.[13] This matters because Ptolemy's research program is not Pythagorean in the loose and popular sense: it does not purchase numbers at the cost of experimental evidence, and in particular at the cost of what the ear reports; on the contrary, it treats the empirical reports of the ear and the mathematical requirements of the mind as joint partners in a fruitful and interesting way[14]—indeed, in such a way as to invite experimentation. I'm not sure that Zarlino had sufficient scientific talent or interest to carry out the Ptolemaic

---

[11] This is the picture of Zarlino that comes into especially sharp focus in the papers by Palisca and Drake cited above. The title of the work in which Vincenzo presented his findings was "Discorso *intorno all'opere di Gioseffo Zarlino*": it was published in Florence in 1589.

[12] Drake's evidence for this claim is pretty compelling. He considers the law for the equilibrium of bodies on the inclined plane. He points out that, by 1577, there were no less than three different, conflicting statements of this law circulating in print. One of these statements was true, that of Jordanus Nemorarius, as published by Tartaglia in 1546. The others were naturally false. From our point of view today, the natural thing to do in the face of this divergence would be to design an experiment to determine which, if any, of these statements is true. But Drake says nobody at the time seems to have done this, even though it would not have been hard to design such an experiment. This leads Drake to say that it was not until Galileo, and the experiment he designed to confirm the law of falling bodies, that anybody used experimentation as a way to confirm or refute a mechanical law stated in advance in mathematical terms.

[13] I should add that I am not saying that this Ptolemizing is the only thing to emphasize in Zarlino's theoretical and scientific commitments. For a very detailed discussion of other aspects of Zarlino's research program, and in particular his conception of formal demonstration in the *Dimostrationi harmoniche* published in Venice in 1571, see John Emil Kelleher's unpublished Columbia University Ph.D. dissertation (1991).

[14] Zarlino explicitly emphasises the joint partnership of reason and the senses in the opening lines of Part Two, chapter 20 of the *Isitutioni harmonice* where he says this (as I translate the passage from page 105 of the 1573 edition): "And since, in music, we use not only reason, but also the senses to judge of sounds and voices, and the one not being in conflict with the other, we have true and perfect knowledge of consonance; wherefore it is now necessary to demonstrate the way of submitting to the judgement of sense every determination made by reason so that we may be certain that sense and reason are in thorough agreement and that the case made by reason not be in vain." There is no reason whatsoever to think that this is a throw-away remark. That Palisca and Drake can represent Zarlino as indifferent to, and even disdainful of, the judgements of hearing proves that they were unwilling to see in the text anything except what they wanted to find there.

research program thoroughly or systematically. But especially if it's true that his expressed commitments didn't carry him very far, I think his limitations, coupled with his unmistakable Ptolemaic propagandizing, may have invited the new musical thinkers—people like this Vincenzo Galilei—to exploit aspects of the Ptolemaic research program, as embodied in Zarlino, seemingly against itself. If that's true, it will turn out that the motor driving progress in music theory, and perhaps also in related scientific fields like the new mechanics, may well have been the two figures who are supposed by Palisca and friends to have acted as a brake.

4.2

Let us now try to spell out a little more fully Zarlino's conception of his research program on his own terms. In particular, we want to see how he himself thought mathematical reason and hearing could contribute to this program. The place to look, I think, is not Zarlino's classic textbook of music theory of 1558, *Le istitutioni harmoniche*, but a polemical book he published later—in 1588—called the *Sopplimenti musicali*. Zarlino was already under pressure from Vincenzo Galilei, who had published a vast criticism of him called the *Dialogue of Ancient and Modern Music* in 1581. (This is to music what Galileo's *Dialogues concerning the Two Chief World Systems* would be to astronomy.) Pressed by Vincenzo, Zarlino is trying in 1588 to state as directly and forcefully as he can what his theoretical outlook is. That makes the *Sopplimenti* a potentially valuable document. For our purposes, Book One, chapter 12 is the place to start, because it lays out some of his position on reason and hearing in a helpful way.

Zarlino explicitly characterizes the power of reason (human, as much as divine) as a kind of cause, namely one that produces beauty and goodness. He says that "reason, considered simply and universally, is that which produces and conserves order and agreement [*convenientia*]" (Zarlino 1588, 35). There is, he says moreover, "a science that encompasses all sciences that belong to reason." This science is mathematics. It follows that mathematics—as the universal science of reason—includes a theoretical understanding of beauty (*speculationi delle cose belle*) as manifested in "order and agreement." Indeed, it looks as if beauty is the privileged object of all mathematical speculation. By implication, mathematics is an evaluative science. It teaches reason to recognize that some of its objects are more beautiful than others and to prefer these most of all. Thus it will turn out that, for Zarlino, reason distinctly prefers the greater excellence in "superparticular" proportions of the form n+1:n whose terms belong to the scenario. But now because mathematical reason is a *cause*, its concern for beauty is not merely theoretical; it seeks the preferred beautiful objects of its theoretical speculation in the real world and there it endeavours to put them on display.

So far Zarlino has been speaking of reason and mathematics in quite general terms. But he goes on to say that the "power of harmony" (*la facoltà Harmonica*) is a special application of mathematical reason. This power is not some mysterious cosmic principle, such as you find in the fragments of the early Pythagorean,

Philolaus, who seems to have treated harmony as the glue that holds the world together. For Zarlino, it is in the first instance that power of mathematical reasoning we use when we do music theory. Moreover, it is that power by which our reason produces order and agreement in the class of things heard. Zarlino says that reason "is the tutor and commander (*Istitutrice & Ordinatrice*) of those things that can be heard in that order to which we especially give the name 'ensemble blending' (*Concento*); for it will discover the measures and proportions together through its speculation (*contemplatione*), put them on display through the handiwork that comes from skill (*Arte*) and also through the consequent experience that belongs to custom and habit" (Zarlino 1588, 35).

This account of the power of harmony leads Zarlino to spell out in a very particular way the role of hearing, its relation to reason and its place among the other senses. He says that reason in its pursuit of the mathematical sciences "uses as servants and ministers the instruments of the highest senses—seeing and hearing—which above all other things are appointed to the service of the ruling part of us, which is the intellect, and to judgment—not merely for the sake of pleasure, but rather all the sooner for the sake of that which is fine [sc. beauty] [*ma più tosto per conto dell'honesto*]." It comes as no surprise to learn that hearing—along with seeing—should be the servant of reason; that was already implied in the idea that reason as power of harmony is the "tutor and commander of things heard." But it is important to see that the relation between reason and its servants—seeing as well as hearing—is anything but antagonistic. That is because the supreme senses are specially adapted for the service they render.

Each of the senses reveals the peculiar differences in the class of sensible things it is privy to. Thus seeing reveals the difference between white and black in things visible; hearing that between high and low in pitched sounds; taste that between bitter and sweet; smell that between the skunky and the sweet-smelling in things whiffed, and so on. But unlike the lower senses, seeing and hearing do not merely attend each to its assigned class of sensible things; they assist each other—and moreover in ways that at the same time assist reason. Zarlino has the following sort of example in mind. An orally expounded argument can sometimes be much easier to follow with visual helps like diagrams (or handouts!); but visual things are sometimes more intelligible to us with the help of articulate sound—as when a diagram or special notation all by itself is mysterious until a teacher verbally explains it to us. Precisely because they assist each other and thereby assist reason as well, Zarlino says that seeing and hearing are "*Fratelli Germani.*" I do not think he means to deny that the lower senses sometimes assist each other (and perhaps also the higher senses too): as, for example, when smell comes to the aid of taste. But he would probably insist that, however much taste and smell assist each other, they never contribute to our intellectual understanding of anything—hence they do not serve reason: at least not in the same privileged way as seeing and hearing. So this is part of what qualifies the higher senses as reason's "servants and ministers." But there is something else to add to the story.

The lower senses can indicate what is pleasant and unpleasant in the differences they reveal to us, but Zarlino says there is a fundamental distinction between the pleasant and the unpleasant, on the one hand, and beauty and ugliness on the other; and, he insists that the latter pair of qualities do not belong to differences in touch, taste or smell. They belong exclusively to things seen and heard: e.g., to shape, to the motions of the heavenly bodies and to musical harmony or concert (*Concento*). This means that the higher senses are specially qualified to serve reason, because they are seeking precisely what reason itself seeks, namely beauty. By implication, beautiful things seen and heard and beautiful mathematical objects investigated by reason just are the same things—though apprehended in different ways.[15] When therefore the higher senses assist each other and thereby jointly assist reason, our understanding and appreciation of beauty gets both deeper and sharper. Zarlino writes, "Hence it is not just because each of these two [sc. higher] senses grasps its own, but rather it is because all the while they almost as it were compete with each other [but as friendly competitors, in the way that loving siblings do! -AL] and together strive for science, doctrine and the investigation of those things perfected by their own reason/proportion that they make progress with respect to the fine [sc. the beautiful] and the useful as well—whence they shine forth; and, those things with a share of reason [sc. the mathematical sciences -AL] prove their worth insofar as they rest upon them [sc. the higher senses -AL]" (Zarlino 1588, 36). It is not just that reason and the higher senses both seek the same object, namely beauty, independently of each other. The point seems to be that they jointly make progress towards this object only because, though reason is their tutor and commander, the higher senses make their peculiar contribution—a contribution that the mathematical sciences cannot do without.

Before pursuing the significance of these reflections further, there are just two final details to clarify in the story. First, we want to understand in what way reason depends on hearing and seeing. This one is easy: Zarlino just means that all reasoning begins with the senses, and here he cites Aristotle as his authority. But secondly, we want to understand in what way reason is the commander and tutor of hearing. That has to mean that hearing is occasionally corrected by reason. This idea we can make both concrete and precise in light of a little melody by Christian Huygens. It proceeds by just intervals accommodated by Zarlino's preferred tuning system, the syntonic diatonic: we go up a perfect fifth from an initial middle C to G, descend a perfect fourth to D, rise another perfect fifth to A, descend by another perfect fourth to E, and then descend a major third to middle C again. But do we really end up where we started from? What does the ear say? As Huygens himself points out, the singer will almost certainly return to the same middle C, because he or she will read middle C in the notation and will give it back to us from memory. But what does reason tell us? Reason's answer depends on working out the ratios. When you do the simple arithmetic, you will discover that the final middle C is sharp by the syntonic comma: 81:80. This means that the singer has faked a pure major third, perhaps

[15] There will be further confirmation of this a little further along—in the specific case of the identity of the consonant intervals and the superparticular proportions whose terms do not exceed six.

without being aware of it. In any case, the fudging is detected by reason, even if it is overlooked by the ear. (In fact, if the series of intervals is repeated n times, reason predicts that the error should ideally accumulate that many times over.) Something like this is what Zarlino has to have in mind when he says that reason tutors or corrects the ear.[16]

Now the ideas that I just finished commenting on from the *Sopplimenti* are not original to Zarlino. They come from Ptolemy. Our clue to their origin is that Zarlino opens his reflections in this chapter by explicitly mentioning Ptolemy's account of the "power of harmony" and what class of power it may be. Good Renaissance humanist though he is, Zarlino uncharacteristically neglects to give us the reference. So I will tell you that this account happens to be the subject of chapter 3, Book Three of Ptolemy's treatise on music—the *Harmonics*.[17] When you read that chapter, you discover that Zarlino has not simply paraphrased Ptolemy; he has systematically translated Ptolemy's Greek—or perhaps the Latin of Antonio Gogava's translation

---

[16]Now some readers may find it paradoxical that I represent Zarlino as hypothetically responding to the Huygens melody in this way, i.e., not defensively at all, but by hypothetically using it to illustrate his idea that reason occasionally has to correct the senses in a branch of mixed mathematics like music theory. For the Huygens melody very naturally recalls the much more complicated, polyphonic examples engineered by Giovanni Benedetti to illustrate the same phenomenon, namely that of drift in pitch in the performance of music written in the just tuning system of Zarlino. These examples can be found in a couple of letters to the composer, Cypriano Rore, that are believed to have been written around 1563 and that were subsequently printed in Benedetti (1585). Palisca argued in "Scientific Empiricism in Musical Thought" that Benedetti proved by means of these examples that Zarlino's tuning system was practically unviable and that it was necessary to adopt some kind of tempered tuning system. But, as it stands, this claim is overstated. The examples prove only that some kind of deviation from Zarlino's tuning system is required if one insisted, for example, on ending the Huygens melody and its more complicated variants on the same pitch one started with. Benedetti himself is explicit about this in the opening lines of the second letter, as Palisca himself notes (1961, 118). But suppose one just doesn't care if there is drift in the performance? Zarlino can be understood as responding to Palisca by exploiting this rhetorical question in the following way. As Benedetti himself points out, reason predicts that this drift will in fact sometimes take place. If the ear does not detect it, that just proves that the ear needs to be corrected by reason, so as to get the truth right, but then, in that case, there is no harm in allowing the drift to take place for the purposes of performance. Now one might well prefer to prevent the drift from taking place, precisely so as not to offend the ear, as when the singing voices are accompanied by instruments that, because of the physical constraints of their construction, don't drift freely in the same way. But since Zarlino was aware of the physical constraints of instrument design and construction, he could have perfectly well conceded that some kind of deviation would indeed be required for such performances without undermining his position in any way. For all such a case proves is that the human voice is freer to drift in the way predicted by reason than certain musical instruments. There would be nothing incoherent, paradoxical, or self-defeating if Zarlino hypothetically made such a move. Indeed, his theoretical commitments are precisely such as to make this move natural and inviting to him. I am indebted to Peter Schubert for helping me sharpen this point.

[17]I should say that, having recognized the Zarlino passage as having been lifted from Ptolemy, I have been influenced in my reading of it by the account that Barker gives of the original Ptolemy passage in his *Scientific Method in Ptolemy's 'Harmonics'* (2000)—see the last chapter of this book. This is because Barker just seems to get Ptolemy's text right; to that extent, I think his account can help shed light on what Zarlino is saying.

of 1562[18]—into Italian: pretty much word for word.[19] ("Power of harmony" is my English for Zarlino's "*facoltà Harmonica*," which is Zarlino's Italian for Ptolemy's "*dynamis harmonike.*") This could just mean that Zarlino was a plagiarist—though that would be fitting, if so, since Porphyry pretty much levels the same charge against Ptolemy in his *Commentary* on the *Harmonics* (5.7–18). But I suspect that plagiarism doesn't really explain what's going on here. Zarlino's appropriation of Ptolemy's words may well reflect his sense that these words are absolutely true; and, it may express his willingness to stand by them just as if they were words of his own.[20] If that's right, Zarlino is signing on to more than the syntonic diatonic scale Ptolemy worked out in the *Harmonics*; he is signing on to the whole Ptolemaic research program in musical science.[21]

Thus it is interesting, and possibly significant, to note that when Zarlino states the object or aim of music theory two chapters earlier in the *Sopplimenti*, he again helps himself to Ptolemy's words in Book One, chapter 2 of the *Harmonics*—again without reference or any kind of acknowledgement. Thus he writes:

> The aim or object of that musical science called ... harmonics is none other than to wish to defend, conserve, and demonstrate with reason the things rationally posited or the rational proportions [*le Positioni ò Proportioni rationali*] of the ... canon or rule, these rationally posited things in no way being in conflict with sensation, according to most people's opinion—just as it is also the aim or object of astronomy to conserve the things posited in harmony with the [observed – AL] motions of the heavens [*le positioni consonanti de i Moti celesti*] .... Since the peculiar concern of speculation or contemplation is to demonstrate that the works of nature are made with reason and ordered causes and that nothing in these works is made preposterously or randomly, especially in those constructions that are the two most beautiful, most worthy, most honoured and most useful of all, those are the most rational senses: seeing and hearing .... (Zarlino 1588, 31)

The passage from Ptolemy Book One, chapter 2 Zarlino has appropriated reads as follows in Andrew Barker's translation:

> The aim of the student of Harmonics must be to preserve in all respects the rational postulates [*hypotheseis*] of the *kanon*, as never in any way conflicting with the perceptions that correspond to most people's estimation, just as the astronomer's aim is to preserve the

[18]Palisca says without qualification that Zarlino did not read Greek. Without qualification, this can't be true, since Zarlino routinely quotes from Greek sources in Greek. But perhaps his command of Greek was not strong; perhaps he was more comfortable with Latin. Palisca mentions that Gogava's translation was available in 1562. See his introduction to Girolamo Mei (1519–1594), *Letters on Ancient and Modern Music to Vincenzo Galilei and Giovanni Bardi* (1977, 41).

[19]This observation raises the question whether there are departures in Zarlino's Italian from Ptolemy's original and, if so, whether these departures are significant. There are a few: (1) seeing and hearing are "*fratelli germani*" rather than sisters; (2) they quasi compete with another; (3) "*concento*" for "*emmeleia.*"

[20]This is strongly suggested by the fact that the opening lines of Part Two, Chapter 20 of the *Institutioni harmoniche* briefly but explicitly state the main point of the Ptolemaic position, namely that reason and the senses are co-judges in music theory, in a voice that is unmistakably recognizable as Zarlino's own.

[21]It is worth noting in this connection the relation between Vincenzo's *Dialogo della musica antica & moderna* and Mei's letters. Vincenzo seems to appropriate key points as his own from Mei without attribution. See Palisca (1977).

> postulates concerning the movements of the heavenly bodies in concord with their carefully observed courses, these postulates themselves having been taken from the obvious and rough and ready phenomena, but finding the points of detail as accurately as possible through reason. For in everything it is the proper task of the theoretical scientist to show that the works of nature are crafted with reason and with an orderly cause, and that nothing is produced by nature at random or just anyhow, especially in its most beautiful constructions, the kinds that belong to the more rational of the senses, sight and hearing.[22]

Let's focus on Ptolemy's own words, which are less muddy than Zarlino's (either because I just haven't rendered them successfully into English, or because Zarlino [or perhaps a Latin source he is using] hasn't successfully rendered Ptolemy into Italian—or both!). Notice that Ptolemy insists—and Zarlino dutifully follows in this—that the student of musical science seeks to reconcile what most people hear as beautiful with the "rational postulates," i.e., the mathematical theory of music.

Now we saw right at the beginning of my paper that Zarlino conceived of music theory as a kind of mixed mathematics. We can now see, in light of his borrowings from Ptolemy, that the conception of music and mixed mathematics he has signed on to is a very particular one. Ptolemy himself quite self-consciously rejects the rival conception—the one he usually calls "Pythagorean"—which we find articulated, for example, in Book VII of Plato's *Republic*. What's distinctive about the so-called "Pythagorean" approach is its sometimes frank scorn for hearing. Here Socrates' criticism in *Republic* VII of the music theory of Plato's day is instructive:

> Don't you realize that in harmonics [sc. as actually studied in Plato's day] they do something . . . like this: they measure heard concords and notes against one another, and so labor to no purpose . . . . "Yes, by the gods," he [Glaucon] said. "Their behavior is quite ridiculous, when they name some things '*pyknomata*' and incline their ears as if hunting out a sound from next door, some of them asserting that they can still just hear a sound in between, and that that is the smallest interval, by which measurement is to be made, while others take issue with them, saying that the notes sounded are already the same, each group putting their ears ahead of their mind." (531a)[23]

The fundamental complaint here is that the music theorists of Plato's day "put their ears ahead of their mind." It is reasonable to ask what, on Plato's view, a music theorist *should* investigate, if not the notes and concords we actually hear. The answer of *Republic* VII is: notes and concords intelligible to the mind. If we ask what such a music theory would be like, the answer may now lie in the music treatise attributed (perhaps wrongly) to Euclid, another text in the tradition Ptolemy would call "Pythagorean." This text, the so-called *Sectio canonis* tries, among other things, to prove formally—in the style of a mathematical treatise—that intervals between notes of different pitch should be treated as numerical ratios and that the ratios corresponding to the consonant intervals are either multiple, i.e., of the form nm:n, or epimoric, i.e., of the form n+1:n. The only concession the author makes to hearing is to accept as consonant intervals the octave (2:1), the fifth (3:2), and

[22]See Barker (1989, 278–278). For the Greek text, see Düring (1930).

[23]As translated by Andrew Barker (1989, 55–56). For the Greek text, see Burnet (1992).

the fourth (4:3). But the author passes over in embarrassed or contemptuous silence the octave plus a fourth, which most ears willing to accept the simple fourth are happy enough to admit as a concord. Why does he neglect this interval? Presumably because the ratio that corresponds to it—(2:1) (4:3)—is 8:3, a ratio which is neither multiple nor epimoric. But who cares what the ears think about the octave plus a fourth?[24]

Ptolemy cares. When he says that the student of harmonics must seek agreement between hearing and reason, he means not that hearing has to accommodate itself to reason's preferred postulates, but rather that reason must begin with the ear's pronouncements and accommodate itself to them. A music theory that jettisons the octave and a fourth over the protest of my ear just isn't credible.

At the same time, Ptolemy is not prepared to side against the Pythagoreans with a quite different rival in music theory—Aristoxenus, a student of Aristotle, who argued, in effect, that since the ear hears musical intervals in the peculiar way it does, and *not* as numerical ratios, the numerical ratios and what follows from them mathematically are irrelevant for the science of music.[25] Aristoxenus, or anyway his disciples, had been understood to say that the ear is always the first and final arbiter in musical debates. This is unacceptable to Ptolemy. For one thing, though the ear doesn't apprehend musical phenomena as numerical ratios, reason does: reason and hearing apprehend the same phenomena—just under different aspects. That leads Ptolemy to say—against both the Pythagoreans and Aristoxenus, as he understands them—that the criterion of truth for settling disputes in music theory is neither reason alone, nor hearing alone, but both together, each in its own special way. Precisely because reason and hearing together serve as criteria of truth, there is hope of achieving the aim of harmonic science, namely to preserve the postulates of reason—its commitment to the relevant numerical ratios—so as to agree with what most people hear. Zarlino follows Ptolemy as faithfully and as literally in this as he has in everything else we have seen.[26]

In a research program like that of Ptolemy and Zarlino, experimentation has to play a central role. It's the only way that reason can assert its authority as a criterion and justify its right to correct the imprecision of hearing in cases like that of the little melody by Huygens we discussed earlier. (Ptolemy would almost certainly have interpreted its significance as I suggested.) Ptolemy never loses sight of this implication. Indeed, he discusses at length different experimental strategies and different experimental devices for demonstrating reason's theoretical commitment to the ear's satisfaction. The devices he considers include: the one string "kanon" or monochord, the eight-string "kanon," the helikon, an extension of the helikon used to test divisions of the tetrachord (a four note scalar substructure spanning a fourth).

---

[24]For the Greek text of this work see von Jan (1895).

[25]For the Greek text of Aristoxenus's *Harmonics*, see Henry Macran (1902).

[26]Here compare the words of Ptolemy's *Harmonics*, Book One, chapter 1 with Zarlino's *Sopplimenti*, Book One, chapter 13.

Ptolemy is not merely interested in the geometry of these devices; he seems genuinely interested in solving problems in their construction. This has suggested to Andrew Barker recently that, even if Ptolemy never built and used these devices, his accounts of them are intended to invite his readers to try them out (Barker 2000, 192–230).

What about experimentation in Zarlino? The implications are the same for Zarlino as they are for Ptolemy. Zarlino certainly talks about the monochord and describes it, as Ptolemy does, as the instrument reason uses to reconcile itself with hearing.[27] He does not, so far as I know right now, examine in any detail the questions about its construction that Ptolemy considers. Nor does he seem to mention devices other than the one-string monochord. On the other hand, his ear recognized the difference between Pythagorean thirds and the pure major thirds of Renaissance polyphony; and, he does officially say that the monochord demonstrates that the pure major third corresponds to the numerical ratio 5:4.[28] He's right about that. To this extent, he is not merely accepting Ptolemy's syntonic diatonic as a tuning system; he is putting into practice Ptolemy's research program in music theory.

4.3

If I am right about the relation between Ptolemy and Zarlino, debates in early modern music theory—some of them, anyway—must be reassessed. They should not be understood as a triumph of modern experimental science over blind superstition. That story is neither fair nor interesting as an explanation of what went on between Zarlino and critics like Vincenzo Galilei. The story is much more complex, and richer philosophically than Palisca and friends imagined. However limited he might have been in certain ways, I would like to suggest that Zarlino invited music theorists—but not just music theorists, perhaps others as well, as I'll go on suggest—to think through the implications of a powerful research program and in effect gave some of them a strategy for moving forward.

The example of Vincenzo tends to confirm my suggestion. The *Dialogue on Ancient and Modern Music* is such a vast, rambling enterprise that it may not make too much sense to describe it as having some one argumentative strategy. But there are fairly clear signs along the way that at least part of Vincenzo's plan of attack is to turn the Ptolemaic research program against Zarlino. The central question of the dialogue is whether the music of the day is sung in the syntonic diatonic of Ptolemy, as Zarlino claims, or in some other diatonic tuning system. Vincenzo's characters in the dialogues, Bardi and Strozzi, are out to show that Zarlino got even this much wrong. But Bardi says explicitly that they will make their case using precisely the same principles as their opponent (Galilei 1581, 4). Bardi and Strozzi then systematically review all the numerical ratios associated

---

[27] Thus, see Part Two, chapter 27 of the *Institutioni harmoniche*.

[28] See Part Two, chapters 39 and 40 of the *Institutioni harmoniche*.

with the intervals of the Ptolemaic syntonic diatonic, the order of the major and minor tones and the semitones in the scale associated with this tuning system, and then they work out the system's theoretical implications on purely rational, i.e., mathematical grounds, by "adding" and "subtracting" the relevant ratios. They try to show as a result of this analysis that a commitment to the Ptolemaic syntonic diatonic would lead to unwanted dissonances; then they try to show experimentally that this is in fact the case. As I have briefly described it, even without the details, the exercise clearly presupposes Ptolemy's twin criteria of truth in music theory: mathematical reason and hearing. Its express aim is to show that these criteria jointly testify against Zarlino's important claim about contemporary musical practice. If this is the right way to characterize the main argumentative strategy of the *Dialogue*, then there is a better way of understanding Vincenzo's relationship to Zarlino than that suggested by Palisco and friends. This is not a story of the triumph of a newly emergent modern scientific way of looking at things over superstitious "Pythagorean" number mysticism. If there was any kind of victory here, it was that of a powerful research program in music theory over a more limited application of itself.

Let us now turn to Galileo and mechanics. Stillman Drake has tried to argue that there is an historical connection between Galileo's discovery of the new mechanics and the debates in music theory we have been discussing. His way of accounting for this connection is to raise a very interesting historical question and to offer a problematic answer by way of his vexed understanding of Vincenzo and his relation to Zarlino.

In Day Three of the *Discourses*, there is a report of a famous experiment Galileo is supposed to have performed on inclined planes to confirm the law of falling bodies, i.e., the law according to which the distance through which a body falls uniformly from rest is directly proportional to the square of the time. Scholars such as Alexandre Koyré used to doubt that Galileo had really performed this experiment. But Thomas Settle showed in 1961 that not only could Galileo have performed the experiment with the technology of his time, he could have done so with an astonishing degree of accuracy (Settle 1961, 19–23). If Drake is right, there is no precedent for an experiment like this in the history of mechanics before Galileo, which raises his interesting historical question: how was Galileo moved to seek experimental confirmation for mechanical laws like that of falling bodies? Drake wants to say that the answer may well lie in Vincenzo's appeal to experiment in his disputes with Zarlino—perhaps especially the experiment I mentioned in Section Two that in effect debunks the legend reported by Nicomachus about Pythagoras and the blacksmith. But this suggestion is problematic *at least* to the extent that it represents Vincenzo on the side of enlightened science and Zarlino on the side of reactionary Pythagorean superstition—where "Pythagorean" is taken in the loose and popular sense I spelled out earlier. If I have got it right, Vincenzo and Zarlino were not facing off on opposite sides; perhaps what made their dispute so bitter—besides the obviously personal sneering and sniping—is that they stood so firmly on the same side, and Vincenzo sought to radicalize it from within. If, then, Galileo found an example in his father, perhaps it was that of an articulate spokesman for the Ptolemaic research program; perhaps Vincenzo—among others[29]—gave him the idea that this research program could be profitably applied to mechanics, indeed in such a way as to make mechanics a branch of mixed mathematics on the same footing as music theory. One would naturally expect mechanics should count as "mixed mathematics," if anything does: it can be understood to "mix" mathematics with a particular class of visible phenomena—the motions of terrestrial bodies. Indeed, from our point of view, it succeeded at this mix far more brilliantly than music theory ever did for the phenomena of hearing. But mechanics did not originally have a place in the Pythagorean curriculum along side music theory and astronomy. In fact, one way of understanding Galileo's achievement is to say that he, more than any one, made a home for it there.[30] If that's the right way to think of his achievement, then it does seem natural to wonder whether Vincenzo (or others) suggested by his (or their) example that Ptolemy had the right approach to mixed mathematics, even if he (Ptolemy) got the system of the world all wrong.

My remarks about Galileo are highly conjectural, and must ultimately remain so. But one thing that supports them, I think, is that the discussion of uniformly accelerated motion in Day Three of the *Discourses* seems to presuppose Ptolemy's twin criteria of truth—mathematical reasoning and the higher senses[31]—even more straightforwardly than the discussion in Vincenzo's dialogue.

At the beginning of this discussion, Galileo presents two at least verbally different definitions of uniformly accelerated motion. His own preferred definition is that such motion acquires equal increments of speed in equal increments of time. But he recognizes that it seems more natural to define it not in terms of time, but in terms of the distance through which a body falls from rest, because it just seems obvious that the greater the distance through which you let a body fall, the greater will be its final speed. Having put these two definitions on the table, Galileo tries to show that the difference between them is not merely verbal, i.e., that they have very different mathematical consequences. To this end, he shows first that the definition in terms of distance implies instantaneous motion, i.e., the absurdity of a motion through a determinate distance in no time at all. Then he shows that the definition in terms of time implies the so-called "Mean-Speed Theorem," which in turn implies the famous law of falling bodies I stated above. Thus the mathematical consequences of the two definitions are very different indeed: the consequences of the one are absurd, while those of the other are not. In effect, Galileo has subjected the two definitions of uniformly accelerated motion to the first of Ptolemy's criteria of truth—that of

---

[29] I have a hard time believing that Vincenzo could have been the sole source of this idea for Galileo. I would want to see more of the historical and intellectual context than I have so far before pronouncing on this.

[30] Perhaps I should be more specific here and speak of dynamics in connection with Galileo's achievement, since it might be argued that Archimedes—and the tradition coming after him—had made statics a branch of mixed mathematics. And surely we might reasonably include statics as a part of mechanics.

[31] Especially seeing, of course, but hearing too, judging from Settle's reconstruction of the experiment on inclined planes (1961).

mathematical reasoning. He has shown that this criterion rejects the definition in terms of distance in favor of that in terms of time. But Galileo does not leave it at that; he also requires some kind of evidence that uniformly accelerated motion, as characterized by his definition in terms of time, is really observed in nature. The next step is, therefore in effect, to subject the definition to Ptolemy's second criterion of truth, i.e., the higher senses. Just as Ptolemy designs experimental devices to test the division of the octave and the divisions of the tetrachord according to "rational postulates," i.e., to verify that the ear actually hears the intervals specified by the rational postulates, so Galileo designs an experimental apparatus (the inclined plane) to test the law of falling bodies, i.e., he finds a way to make the higher senses testify that the consequence of his preferred definition of accelerated motion actually obtains in nature.

What this goes to show is that it seems very natural to read the discussion of accelerated motion in Day Three of the *Discourses* as a novel implementation of the Ptolemaic research program—not in music theory this time, but in the newly emerging science of mechanics. It leaves totally untouched the question of Galileo's self-understanding: whether Galileo would have recognized himself as having signed on to the Ptolemaic research program or not. At the same time, it raises a larger historical question whether and how Ptolemy's approach to mixed mathematics—whether in music theory, astronomy or optics—might have had an impact on the emergence of mechanics as a branch of mixed mathematics in the early modern period. But that's a question that might well be worthpursuing.

References

Barker, Andrew. 1989. *Greek musical writings*. Vol. 2. Cambridge: Cambridge University Press.

Barker, Andrew. 2000. *Scientific method in Ptolemy's "Harmonics"*. Cambridge: Cambridge University Press.

Benedetti, Giovanni. 1585. *Diversarum speculationum mathematicarum et physicarum liber*. Tuarini, apud haeredem Nicolai Bevilaque.

Burnet, John, ed. 1992. *Platonis Opera*. Vol. 4. Oxford: Oxford University Press.

Cohen, H. F. 1984. *Quantifying music: The science of music at the first stage of the scientific revolution, 1580–1650*. Dordrecht: De Reidel Publishing Company.

Diels and Kranz. 1989. *Die Fragmente der Vorsokratiker*. Vol. 1. Zürich: Weidmann.

Drake, Stillman. 1970. *Galileo studies*. Ann Arbor, MI: The University of Michigan Press.

Düring, Ingemar, ed. 1930. *Die Harmonielehre des Klaudios Ptolemaios*. Göteborg: Elanders Boktryckeri Aktiebolag.

Düring, Ingemar, ed. 1932. *Porphyrios Kommentar zur Harmonielehre des Ptolemaios*. Göteborg: Elanders Boktryckeri Aktibolag.

Galilei, Vincenzo. 1581. *Dialogo della musica antica et della moderna*. Florence: Appresso Giorgio Marescotti.

Kelleher, John Emil. 1991. *Zarlino's "Dimostrationi Harmoniche" and demonstrative methodologies in the sixteenth century*. PhD diss., Columbia University.

Macran, Henry, ed. 1902. *The "Harmonics" of Aristoxenus*. Oxford: Oxford University Press.

Mersenne, Marin. 1666. *L'Harmonie universelle*. Livre premier des consonances. Facsimle repr. Paris: Éditions du Centre national de la recherche, 1965.

Palisca, Claude. 1961. Scientific empiricism in musical thought. In *Seventeenth-century science and the arts*, ed. Hedley Howell Rhys, 91–137. Princeton, NJ: Princeton University Press.
Palisca, Claude. 1977. Introduction to *Letters on ancient and modern music to Vincenzo Galilei and Giovanni Bardi*, by Girolamo Mei. Second edition. American Institute of Music. Hänssler Verlag: Neuhausen-Stuttgart.
Settle, Thomas. 1961. An experiment in the history of science. *Science* 133: 19–23.
von Jan, Karl, ed. 1895. *Musici scriptores graeci*. Teubner: Leipzig.
Zarlino, Gioseffo. 1573. *Le istitutioni harmoniche*. Venice: Appresso Francesco de I Franceschi Senese. Repr. Ridgewoord, NJ: Gregg Press Incorporated, 1966.
Zarlino, Gioseffo. 1588. *Sopplimenti musicali*. Venice: Francesco de' Franceschi, Sanese.

Part II
Creating Traditions

Chapter 5
Ethics in Descartes and Seventeenth Century Cartesian Textbooks

Roger Ariew

Descartes wrote the *Principles of Philosophy* as something of a rival to scholastic textbooks. He prided himself in "that those who have not yet learned the philosophy of the schools will learn it more easily from this book than from their teachers, because by the same means they will learn to scorn it, and even the most mediocre teachers will be capable of teaching my philosophy by means of this book alone" (AT III, 259–60). Still, what Descartes produced was inadequate for the task. The topics of scholastic textbooks ranged much more broadly than those of Descartes's *Principles*; they usually had quadripartite arrangements mirroring the structure of the collegiate curriculum, divided as they typically were into logic, ethics, physics, and metaphysics.[1] But Descartes produced at best only what could be called a general metaphysics and a partial physics. Knowing what a scholastic course in physics would look like, Descartes understood that he needed to write at least two further parts to his *Principles of Philosophy*: "a fifth part on living things, i.e. animals and plants, and a sixth part on man"; he did not do so, as he said, because "I am not yet completely clear about all of the matters I would like to treat in these last two parts, and do not know whether I am likely to have sufficient leisure <or be able to make the experiments necessary> to complete them" (*Principles* IV, art. 188). And he did not issue what would be called a particular metaphysics.[2] Descartes, of course, saw himself as presenting Cartesian metaphysics as well as physics, both the roots and trunk of his tree of philosophy. But from the point of view of school texts, the metaphysical elements of physics (general metaphysics) that Descartes discussed—such as the principles of bodies: matter, form, and privation; causation; motion: generation and corruption, growth and diminution; place, void, infinity, and time—were usually taught at the beginning of the course on physics. The scholastic course on metaphysics—particular metaphysics—dealt with other topics, not

R. Ariew (✉)
Department of Philosophy, University of South Florida, Tampa, FL, USA
e-mail: rariew@cas.usf.edu

[1] For the collegiate curriculum in seventeenth-century France, see Brockliss (1987).

[2] As Marion argues (1986, 9–72), Descartes did not produce a metaphysics of being *qua* being. See also Ariew (1999).

C. Fraenkel et al. (eds.), *The Rationalists: Between Tradition and Innovation*,
The New Synthese Historical Library 65, DOI 10.1007/978-90-481-9385-1_5,

discussed directly in the *Principles*, such as: being, existence, and essence; unity, quantity, and individuation; truth and falsity; good and evil. Such courses usually ended up with questions about knowledge of God, names or attributes of God, God's will and power, and God's goodness. Thus the *Principles of Philosophy* by itself was not sufficient as a text for the standard course in metaphysics. And, of course, Descartes also did not produce a text in ethics or logic for his followers to use or to teach from.

These must have been perceived as glaring deficiencies in the Cartesian program and in the aspiration to replace Aristotelian philosophy in the schools. So the Cartesians rushed in to fill the voids. One could mention their attempts to complete the physics—Louis de la Forge's additions to the *Traité de l'homme*, for example—or to produce more conventional-looking metaphysics—such as Johann Clauberg's later editions of his *Ontosophia* or Spinoza's *Cogitata metaphysica.* In the *Ontosophia*, Clauberg discusses being in general, dividing it into three: its general and primary sense of "intelligible" being, a secondary and lesser sense of "something" to be distinguished from "nothing," and a third, particular sense of "real" being, being outside the intellect, or substance, as opposed to accident and mode. Clauberg goes on to talk about essence, existence, and duration. His remaining chapters concern pairs of concepts such as one and many, true and false, good and evil, perfect and imperfect, distinct and opposite, the same and another, exemplar and image. Similarly, in the *Cogitata metaphysica* Spinoza discusses beings: real beings, fictitious beings, and beings of reason; he then proceeds to the being of essence, existence, idea, and power. After chapters on necessary, impossible, possible, and contingent, he produces ones on duration and time, order, and the transcendental predicates: the one, the true, and the good. In Part II Spinoza discusses, one after another, God's eternity, his unity, immensity, immutability, simplicity, life, intellect, will, power, creation, and concurrence, before ending with a chapter on the human mind. Again, these are standard topics in scholastic texts on metaphysics; whether all of them can be dealt with in Cartesian philosophy is a matter of disputation, but the fact is that they were not dealt with directly in Descartes's *Principles.* Cartesians in the seventeenth century also began to produce the kinds of texts not normally associated with their intellectual movement, that is treatises on ethics and logic, the most prominent of the latter being the *Port-Royal Logic* (Paris, 1662).

The attempt to publish a Cartesian textbook that would mirror what was taught in the schools culminated in the famous multi-volume works of Pierre-Sylvain Régis and of Antoine Le Grand. The Franciscan monk Le Grand initially published a popular version of Descartes's philosophy in the form of a scholastic textbook, expanding it in the 1670s and 1680s[3]; the work, *Institution of Philosophy,* was then translated into English together with other texts of Le Grand and published as *An Entire Body of Philosophy according to the Principles of the famous Renate Descartes* (London,

---

[3] *Philosophia veterum e mente Renati Descartes, more scholastica breviter digesta* (London, 1671) and *Institutio philosophia, secundum principia Renati Descartes, nova methodo adornata et explicata ad usum juventutis academicae* (London, 1672, 1678, and 1680).

1694). On the Continent, Régis issued his *Système Général selon les Principes de Descartes* at about the same time (Amsterdam, 1691), having had difficulties receiving permission to publish. The *Système Général*, the great complete Cartesian textbook, is a very odd work. Although billed as a general system of Cartesian philosophy, it does not seem very systematic (in our sense of the word). Its various portions embody Régis's adaptations of diverse philosophies, both Cartesian and non-Cartesian: Arnauld's "*Port-Royal*" logic (mostly excerpted, though with some changes); Robert Desgabet's peculiar metaphysics[4]; Jacques Rohault's physics; and an amalgam of Gassendist and Hobbesian ethics.[5] Ultimately, Régis's unsystematic (and very often un-Cartesian) *Système* set the standard for Cartesian textbooks.

There were other attempts at setting out a complete Cartesian system before those of Le Grand and Régis. The first such work I know of is Jacques Du Roure's *La Philosophie divisée en toutes ses parties* (Paris, 1654) and its successor *Abrégé de la vraye philosophie* (Paris, 1665). Du Roure was one of the first followers of Descartes, belonging to the group around Descartes's literary executor, Claude Clerselier. In Du Roure's case, the parts of philosophy included natural theology plus the usual parts of the curriculum. Thus, Du Roure was also the first to have written a Cartesian ethics and natural theology. Du Roure and Clauberg, who apparently made each other's acquaintance during the latter's trip to France in 1648, wrote the first Cartesian-style logic texts.[6]

Perhaps the most interesting attempt at a Cartesian ethics is the Latin-language student manual on Descartes's ethical thought printed in London in 1685.[7] Descartes never wrote such a book, but the clever translator was able to put together a three-part treatise out of Descartes's own words from his correspondence with Christina, Elisabeth, Mesland, and Chanut, and from *The Passions of the Soul*. Apparently, this manual became part of the curriculum at Cambridge University since it was republished numerous times there during the first three or four decades of the eighteenth century and bound into a single volume together with the "scholastic" ethics of Eustachius a Sancto Paulo (that is, partII of his *Summa philosophiae quadripartita*)

[4]For an account of the peculiarities of the Cartesian metaphysics of Desgabets and Régis, see Schmaltz (2002).

[5]A contemporary description of the work, from a letter by Simon Foucher to Leibniz, confirms this impression of eclecticism: "You know that I think Regis has given the public a great system of philosophy in 3 quarto volumes with several figures. This work contains many very important treatises, such as the one on percussion by Mariotte, chemistry by l'Emeri, medicine by Vieuxsang and by d'Uvernai. He even speaks of my treatise on Hygrometers, although he does not name it. There is in it a good portion of the physics of Rohault and he refutes there Malbranche, Perraut, Varignon—the first concerning ideas, the second concerning weight, and the third, who has recently been received by the Académie royale des Sciences, also concerning weight. The *Metheores* of l'Ami also in part adorn this work, and the remainder is from Descartes. Regis conducted himself rather skillfully in his system, especially in his ethics" (GP I, 398–400).

[6]*Logica vetus et nova* (Amsterdam, 1654 and 1658), and *logica contracta* in *Opera omnia philosophica* (Amsterdam, 1691). I treat the issue of Cartesian Logic in Ariew (2006).

[7]*Ethice. In Methodum et Compendium, Gratiâ Studiosae juventutis, Concinnata.*

and the "Christian" ethics of Etienne de Courcelles (*Synopsis ethices*).[8] This is the main text I wish to consider. Here is its structure, in slightly more detail. Chapter I contains three parts:

1. *De Summo Bono* (from a letter to Christina, 20 November 1647, V, 82–85: C 1)
2. De Vita Beata (from the following letters to Elisabeth: 4 August 1645, IV, 264–266: C 4; 1 September 1645, IV, 281–287: C 6; 15 September 1645, IV, 291–96: C 7; January 1646, IV, 354–356: C 10)
3. *De Libero Arbitrio* (from the following letters to Mesland: 27 May 1641, III, 378–380: C 112.[9] 2 May 1644, IV, 117: C 115; and to Elisabeth: January 1646, IV, 354: C 10; 3 November 1645, IV, 332: C 7)

Chapter II is an abbreviated, and at times reordered, Latin translation of all three parts of *The Passions of the Soul,* with the physiological passages deleted. And chapter III is a treatise on intellectual love starting with a fragment of a letter to Chanut (1 February 1647, IV, 601–606: C 35), "What is Love?" and continuing with a discussion of the following topics:

- Whether natural light alone teaches us to love God? (To Chanut, 1 February 1647, IV, 607–613: C 35)
- What are the causes that often incite us to love someone in preference to another before we know their worth? (To Chanut, 6 June 1647, V, 56–58: C 36)
- Of the two derangements, which one is worse, the one caused by love or the one caused by hate? (To Chanut, 1 February 1647, IV, 613–617: C 35)
- The Joy of Soul (To Elisabeth, October or November 1646, IV, 530: C 15)
- Whether it is better to be cheerful and content, imagining the goods one possesses to be greater and more valuable than they are than to have more consideration and knowledge, so as to know the right value of both and thus to grow more sad? (To Elisabeth, 6 October 1645, IV, 305–308: C 8)

It may look as though the translator has made a concerted selection from Descartes's correspondence, but in fact he is just following Clerselier's edition of it (or, more precisely, the Latin translation of Clerselier's correspondence published in London and Amsterdam in 1668). It seems clear that Clerselier wasn't just publishing a selection of Descartes's correspondence, but was thinking of constructing Cartesian texts to fill the gaps in the extant Cartesian corpus, starting with ethics. Although in his Preface to the *Correspondence* Clerselier says that there is no order to his collection of letters—"I did not give much consideration to the order and

---

[8]Eustachius a Sancto Paulo, Etienne de Courcelles, and René Descartes, *Ethica, sive, Summa moralis disciplinae, in tres partes divisa* (Cambridge, 1707). Eustachius' part is called *Ethica* and de Courcelle's *Synopsis ethices*. Although the work says that it is divided into three parts, it looks really like a two-part work, Eustachius being the first half and de Courcelles-Descartes the second.

[9]This is a variant of IV 704–706, which is itself possibly a Latin translation of the French translation of IV, 173–175.

sequence of the letters in general"—he begins with the letter to Christina on the supreme good and continues with the letters to Elisabeth on the happy life; this is what the Ethics treatise does as well, in an abbreviated fashion, giving a selection from Clerselier, Letters 1, 4, 6, 7, and 10—in that order. Clerselier is clearer about his other motivations. First, he argues that his collection of Descartes's letters is equivalent to any other of Descartes's writings,[10] even though Descartes might not have thought to publish them, since "one should not fear the public censure of what is written for Princesses and for the most learned people in Europe." According to Clerselier, what is addressed to such people, who are esteemed for their rank, knowledge, or virtue, will assuredly be well-considered and highly polished. Clerselier then asserts that the highest and most useful subject is without doubt the one that Descartes examines in his letter to Queen Christina, namely, the topic of the supreme good, which he had just treated as well in the letters to Princess Elisabeth. Clerselier says:

> Descartes allowed people to see, in these letters, that ethics was one of his most common meditations, and that he was not so powerfully engaged with the consideration of things that happen up in the air, or with the inquiry into the secret paths nature observes in the production of its works here below, such that he failed to reflect frequently on himself, and ... to regulate the actions of his life, following the true reason. ... After this, I do not think that anyone will be able to accuse him of vanity in his studies, as being completely engaged with an inquiry into the empty things of which science fills the mind, instead of those that instruct and perfect man.

Clearly, Clerselier thinks he has good reasons to start his collection of Descartes's correspondence with Descartes's letters to Queen Christina and Princess Elisabeth and his thoughts on ethics: the supreme good and the happy life.

In the letter to Christina, Descartes first delineates what is good in itself (in which case, God is the supreme good) and what is good for another. With what is good for another, there is what is good for us, what belongs to us and is a perfection for a set of humans—that is, the set of all goods for the soul, body, and fortune—and what is good for each in particular. In the latter crucial case this is a "firm will to do well and the contentment produced thereby." The obvious contrast here is the Aristotelian definition of good for man as the end of human actions, leading to the ultimate good as associated with man's function, whether taken naturally, or as the scholastics often do, supernaturally. For Descartes, good is a perfection belonging to us; thus the greatest good cannot be connected to the goods of body and fortune, which do not depend upon us, but rather to the goods of the soul. These, in turn, are associated with knowledge, which can surpass us, and with will, which is in our power. As a result, the supreme good is a "firm and constant resolution to do everything we judge to be best and to use all our power of mind to know these." Descartes says, "this by itself constitutes all the virtues; this alone really deserves all the praise and glory; this alone, finally, produces the greatest and most solid contentment in life."

---

[10]"... je te presente ces lettres avec autant de confiance que Monsieur Descartes a pû faire lui-même ses autres écrits, sçachant qu'elles ne cedent en rien à pas un autre ouvrage que tu ayes pû voir de luy."

In this way, Descartes believes he reconciles Stoicism and Epicureanism, since he thinks the greatest good to be in both virtue (vigor of resolution)—or honor (being deserving of praise)—and pleasure (contentment).

The selection from Descartes's letters to Elisabeth reinforces these various points. Descartes distinguishes between good fortune, which is not in our control, and happiness, which is. He brings out an analogy of a vessel that can be filled to capacity with less liquid than another, because it has lesser capacity. Likewise we can be filled with contentment with what is in our control, like virtue and wisdom, without having to need anything external, like honors, wealth, or health (although these do bring extra contentment, that is, extra "fluid"). Descartes continues, "each person can be content ... as long as he observes three things, to which the three rules of morality that [he] put forward in the *Discourse on Method* are related."[11] At which point he details the three maxims, which presumably are the foundations of the three (or four) rules he had proposed in the *Discourse*.

Instead of obeying the customs of his country, Descartes now proposes to attempt always to use his mind as well as possible to discover what he should do in all the circumstances of life. And instead of being as firm and as decisive in his actions as he could and follow even the most dubious opinions, he proposes to have a firm and constant resolution to execute everything reason advises him without allowing his passions or appetites to divert him. Again he places virtue in the firmness of this resolution. Descartes ends by having one consider that in this manner of conduct, the goods we do not possess are entirely outside our power, which allows us not to desire them. In the *Discourse*, Descartes similarly proposed that we always try to master ourselves rather than fortune, and to change our desire rather than the order of the world. At this point in the *Discourse*, he also proposes a review of all occupations to conclude his moral code. In the later ethics, Descartes instead continues with advice on how to achieve virtue, given that we are not just mind, but a mind united to a body, a composite being prone to illnesses and passions. Moreover, he tackles the subject of our imperfect knowledge and what we should keep in mind in order to be disposed always to judge well: that is, the existence of God, the nature of our souls, and our distinctness from every part of the universe. This latter thought contains the beginning of Descartes's social and political theories in that it distinguishes between the interests of the part and the interests of the whole and advises always to prefer the interests of the whole, of which one is a part, than one's own particular interests.

There is much to be said about even this beginning of ethics, but let me move very briefly to its reception by the Cartesians. Only one of the three "Cartesian" ethics texts I consulted reflected significantly upon these materials. As I said, Régis's ethics was an amalgam of Hobbesian and Gassendist ethics; the Hobbesian part is obvious: Régis divides morality into three parts: natural, civil, and Christian, with natural morality holding in the state of nature, civil morality in the political state, and

---

[11] It may be important to note that the translation in CSMK p. 257 (AT IV, 265) seems defective: "each person can make himself content ... provided he respects three conditions, which are related to the three rules of morality which I put forward in the *Discourse on the Method*."

Christian morality in the state of Christianity. Though this is generally Hobbesian, the direct source is most likely Samuel Pufendorf.[12] Régis asserts that in the state of nature we are driven by our self-preservation, which we love, but that we can rarely preserve ourselves without working with others, so that, for true self-love, you must love your neighbor as you would yourself. Ultimately, you cannot do that without loving God. None of this has much to do with Descartes's views on ethics.

Still, in his *Morale ou La connoissance des devoirs de l'homme*, Book II, On the Duties of Man Considered in Civil Society, Part II, "On the Means of Easily Satisfying the Duties of Civil Society," Régis has a small chapter on the "Supreme Good and Happiness of Man in the State of Nature and in Civil Society" (chapter III, 489–491). There he argues that man's greatest perfection consists in taking pleasure in the supreme good and that the supreme good of man in the state of nature and in civil society consists in "everything that contributes to conserve him by the good use he makes of it while following natural and civil laws" (489). Régis distinguishes between the supreme good and the good in general: the latter is what the soul can love while using its freedom, whether well or not, whereas the former concerns only those things of which the soul actually makes good use. He then states, "Since blessedness is nothing other than the enjoyment of the supreme good, man's blessedness in the state of nature and in civil society consists in the internal contentment that the soul receives from the good use it makes of the things that contribute to its conservation" (489). Rejoining Descartes, Régis insists that this natural and civil beatitude is the only contentment which is entirely in man's power, whereas the goods of body and fortune do not depend at all on this power. Thus the contentment relates wholly to two things alone, namely, to understanding and to will:

> but since it is not in man's power to possess the knowledge he is missing, only man's free will remains as that of which he can absolutely dispose. And it is not possible for him to dispose of it better than when he has a constant resolution to do exactly all the things that contribute to his conservation, following what the natural and civil laws prescribe for him. It is that and that alone which, properly speaking, deserves praise and glory, and from that alone results the greatest and most solid contentment of life. (490)

---

[12] In his extremely popular 1672 treatise, *De officio hominis et civis juxta legem naturalem*, Pufendorf similarly divides duty into three, natural, civil, and Christian: "Therefore it is manifest that from three founts, so to speak, men derive the knowledge of their duty and what in this life they must do, as being morally good, and what not to do, as being morally bad: namely the light of reason, the civil laws and the particular revelation of the divine authority. From the first flow the commonest duties of man, especially those which make him sociable with other men; from the second, the duties of man in so far as he lives subject to a particular and definite State; from the third, the duties of a man who is a Christian. From this three separate studies arise, the first of which is the natural law, common to all nations; the second, the civil law of the single individual States, into which the human race departed. The third is called moral theology in contradistinction to that part of theology which explains what is to be believed [that is, dogmatic theology]" (Pufendorf 1927). I owe the reference to Pufendorf to my colleague, Colin Heydt.

In this way, Régis can capture some of Descartes's ethics, slightly modified, within a framework clearly foreign to it.

Du Roure's *Morale* is not any better at representing Descartes's ethics. Du Roure defines the good as what is suitable for something, as "health is always suitable for a sick person," and says that the true good is either physical or moral, with physical good as what is independent of human will, and moral good as what is within our control. Happiness turns out to be the state in which we enjoy a variety of pleasant good that we can reasonably possess. There are goods of the body and goods of the mind. The latter have to do with knowledge and will. Moral virtue turns out to be knowledge accompanied by the firm will to do well: the virtuous person is the one who does and wills the things that must be done. This is as far as it goes. It is a pale echo of Descartes.

With Le Grand, we finally have a reasonable portrayal of Descartes. Le Grand's discussion, the Tenth Part of the *Institution of Philosophy*, covers many pages and ranges broadly over many topics. After arguing that external goods are not the good of man, Le Grand comes to the main question: What is the highest good of man in this life, and his ultimate end? He considers man in a double state, as a private man or as mankind, the latter of which comprehends all men. The supreme good for all mankind is the concurrence of all perfections of which he is capable, that is, the goods of the soul and body and fortune. But for private man the supreme good is the right use of his reason, which consists in "his having a firm and constant purpose of always doing that, which he judges to be the best." This, of course, is in our power, whereas the goods of body and fortune are not.

The proper use of our two main intellectual faculties also produces a satisfaction of mind. The doctrine is encapsulated in the three things we need to observe, which are said to be the foundation of all ethics. The first is that we "strive to attain the knowledge of what we ought to embrace." The second is that "we stand firm and constant to what we have once resolved upon and purposed." And the third is "that we lay down as unmovable ground and principle, that nothing besides our own thought is in our power." Le Grand concludes "that the natural happiness of man is nothing else but that tranquility or joy of mind, which springs from his possession or enjoyment of the highest good," and further argues that, by this notion of the highest good of man, he concurs with the sentiments of Epicurus and Zeno—which is to understand Descartes very well.

To sum up: Cartesian textbooks in the seventeenth century are a mixture of tradition and innovation. While their form and general contents are usually traditional (scholastic) and non-Cartesian, they achieve their innovations at the margins, in their specific doctrines, but these often deviate from proper Cartesian views. In the case of ethics, the manner in which the Cartesians interpret Descartes's ethics might in part be due to their representations of an obscure text, the *Ethica*, an arrangement of materials taken from disparate letters of Descartes in which he discussed matters of ethics.

References

Ariew, Roger. 1999. Descartes and the late Scholastics on the order of the sciences. In *Philosophy in the sixteenth and seventeenth centuries: Conversations with Aristotle,* eds. C. Blackwell, S. Kusukawa, 350–64. Aldershot: Ashgate.
Ariew, Roger. 2006. Descartes, the first Cartesians, and logic. *Oxford Studies in Early Modern Philosophy* 3: 241–260.
Brockliss, L. W. B. 1987. *French higher education in the seventeenth and eighteenth centuries: A cultural history*. Oxford: Clarendon.
Marion, Jean-Luc. 1986. *Sur le prisme métaphysique de Descartes. Constitution et limites de l'onto-théo-logie*. Paris: P. U. F.
Pufendorf, Samuel 1927. Preface to the benevolent reader, trans. H. F. Wright. In *On the duty of man and citizen*, trans. F. G. Moore. Oxford: Oxford University Press.
Schmaltz, Tad. 2002. *Radical Cartesians*. Cambridge: Cambridge University Press.

Chapter 6
Louis Bourguet and the Model of Organic Bodies

François Duchesneau

The presumptions that underpin the preformationist theories of generation during the first decades of the eighteenth century, the period when Antonio Vallisneri (1661–1730) achieved fame, derive, for the most part, from neocartesian positions in the philosophy of nature: these positions can often be traced back to the theses expounded by Malebranche in his major works, including, among others, *Recherche de la vérité,* (1674–1678) and *Entretiens sur la métaphysique* (1688). Such approaches are certainly metaphysical, but they also have epistemological and methodological implications. From that point of view, although Vallisneri's *Istoria della generazione dell'uomo e degli animali* (1721) contains numerous direct and indirect references to Malebranchian tradition, it would be a stretch to consider that that tradition had sole dominion over the work of this Italian scientist, let alone over the work of his contemporaries. In the context of that era, the models of experimental philosophy—to which, from a methodological perspective, most naturalists subscribed—convey a more reserved, even critical, attitude on the scope and intelligibility of metaphysical postulates. At the turn of the eighteenth century, new explicative approaches came about that conveyed the dissemination and exploitation of epistemological models and norms inherited from experimental philosophy, which Newtonian science extended. Finally, we should not neglect the influence of the architectonic concepts and principles of the system of nature developed by Gottfried Wilhelm Leibniz (1646–1716); these concepts and principles were passed on by adepts prone to adjust them to the constraints of the analysis of complex systems characteristic of vital organization.

The exploration that I outline and that deals with the application and transformation of the Leibnizian model of organic bodies, while it was aimed principally at Vallisneri, needed to be undertaken by taking into account a corpus of documents parallel to the published texts, notably the more informal elements of correspondence. Moreover, we should consider, on the whole of this subject, the influence of the Leibnizian concepts of force, organism, and monad; these concepts are principally found in a few of this German philosopher's published works, including

F. Duchesneau (✉)
Université de Montréal, Montréal, QC, Canada
e-mail: francois.duchesneau@montreal.ca

C. Fraenkel et al. (eds.), *The Rationalists: Between Tradition and Innovation,* The New Synthese Historical Library 65, DOI 10.1007/978-90-481-9385-1_6,

Essais de théodicée (1710), and have been interpreted quite liberally. This influence belongs to original methodological contexts that dealt with the observation of complex, organic, and vital phenomena, difficult to reduce to the micromechanistic models that had inspired the first generation of preformationists, those influenced mostly by Malebranche.

Work of this type is essential in the case of Vallisneri and of the larger context to which belongs the contribution of *Istoria della generazione* to the critical examination of the theories of organized bodies and of generation, just as it is also essential in the case, a few decades later, of the Genevan naturalist Charles Bonnet and of the problematical context to which belong his *Considérations sur les corps organisés* (1762) and his subsequent works.

As a partial contribution to such analyses as yet unrealized, I want to relate how Louis Bourguet (1678–1742), one of the major scientific correspondents of both Leibniz and Vallisneri, conceived of reconciling within a same explicative model quasi-antithetical elements of analysis on the pre-existence of microscopic and submicroscopic germs and on the modes of development by which these produced observable living organisms. I thus propose a preliminary analysis of the more pertinent elements of the pivotal text that is Bourguet's *Lettres philosophiques sur la formation des sels et des crystaux, et sur la génération et le mechanisme organique des plantes et des animaux* (1729).[1]

6.1 The Stakes of the Transition from Vallisneri to Bourguet

A few decades ago, Jacques Roger attempted to describe the overall link that allows us to situate Bourguet as an extension of Vallisneri (Roger 1971, 373–378). Referring to the examination of Chapters 14 and 15 of the second part of *Istoria della generazione*, Roger reminds us that Vallisneri, while determinedly critical of generation by spermatic animalcules, defended with great rigour the thesis of the pre-existence of germs as the only methodologically acceptable thesis on generation. However, he had to acknowledge a certain empirical improbability, even the paradoxical character of this thesis should we consider its associated difficulties, notably concerning the causal explanation of monstrosities and of the transmission of hereditary characteristics of both parental lineages in the case of sexual reproduction.

Vallisneri adheres to a methodology of mechanistic explanation that comes to him from the Galilean tradition characteristic of the school of Marcello Malpighi (1628–1694). From that point of view, he does not belong to the epistemological tradition that denounced the insufficiencies and deficiencies of mechanism to justify by compensation the recourse to specific agents, shaping, adjusting, and transforming the natural order and inserting within it a teleological and providential organization of phenomena. Such were the meaning and role of the "plastic nature," the "spirit

[1] See Bourguet (1729).

of nature," or the "hylarchic principle" postulated by the Cambridge neoplatonists Henry More (1614–1687) and Ralph Cudworth (1617–1688). Subsequently, the animism of Claude Perrault (1608–1680) was based on an analogous methodological approach. The inadequacies of the Cartesian type of mechanism to explain the production and operation of organized bodies allowed these authors to justify a recourse to principles of the animistic type, causal agents of some sort inherent to the natural order, able to direct and correct natural mechanisms so as to produce integrated and functional effects. Vallisneri clearly rejects plastic natures, as per the articles that Jean Le Clerc (1657–1736) published in the *Bibliothèque choisie* from 1703 to 1706, which contributed to redefining them counter to the autosufficiency and the hegemony of mechanical explicative reasons. Furthermore, it seems that Vallisneri rejects them for fundamental reasons analogous to Leibniz's in *Considérations sur les principes de vie et sur les natures plastiques* (GP VI, 539–546).

Vallisneri's point of view is characterized by the inclusion of an integral sphere of mechanical determinations of natural phenomena, including vital phenomena. These determinations depend analytically on motor principles such as gravity, elasticity, and fermentation; this is the only order of forces that can be legitimately invoked to explain the production of any natural phenomenon whatsoever. Furthermore, it seems that only fermentation can be considered as a principle in the case of generation. Yet, the observation of the phenomena that characterize this function does not reveal the operative presence of fermentation processes. In any case, the movements due to fermentation, "being unintelligent and blind, will never fashion an organic body, which requires the greatest prescience, the greatest science, and the greatest ability" (Vallisneri 1733, II, 199b–201a).[2]

It's not surprising that, under these conditions, Vallisneri developed a thesis of a certain harmony that could deploy itself in natural movements by reason of their overall mechanical determinations. This harmony would deal with the ordinary effects of movement but could not satisfactorily account for the formation of organic bodies. The hypothesis evoked is that of a functional ordination of movements that would produce that kind of effect but that would depend on a principle that is transcendent in relation to human intellection, thus exempt from our norms of intelligibility. Searching to identify the operations unique to the diverse parts and to transpose these functions in terms of specific structures and movements conforms to the methodology of mechanistic microstructuralists and to our principles of rationality; but attempting to explain the formation of a living being from the mechanical operation of the parents' genital organs would represent an illegitimate analogical extension of the physical explanation true to our principles of intelligibility. It would be like attempting, under the pretext that the brain is the seat of the intellective functions, to explain these functions by way of physiological processes such as the circulation of animal spirits and the pulsations affecting the membranes and fibres of the cerebral system.

---

[2]Quoted in Roger (1971, 373).

Vallisneri's solution does not consist in the presumption of a transcendent mechanism but in admitting a natural order that, in one aspect, that of innate production, remains irreducible to our structures of intelligibility. The "reign of nature" is comprised of elements of organization that we cannot conceive of as issuing from mechanical intelligibility, according to the clear and distinct ideas that we form about mechanisms. That is the meaning of this oft-cited passage:

> Who could tell me that I cannot cull from my ideas the rule of all that I assert and of all that I repudiate, that I, who do not want to assert or repudiate anything that I do not understand clearly and distinctly, do not have the right to repudiate the formation of animals by mechanism, as long as I do not conceive of the slightest trace of the animal in all mechanical laws? (Vallisneri 1733, II, 203a)[3]

We should note, as does Roger, that the specific laws regulating these organic formations irreducible to mechanism must nevertheless answer to the same epistemological criteria to which answer the laws of inorganic nature: both sets of laws must be "necessary, general, simple, and constant." Vallisneri, according to Roger, notes the infinite diversity and the extreme complexity of living beings as natural productions. He thus refuses to subordinate them to mechanical laws, but only to link them to developmental laws underpinning species constancy. Hence the plausible character from which would benefit the hypothesis of pre-existence to the exclusion of all others. Thus could we reconcile the quasi-infinite diversity of organic productions, stemming from a transcendent teleological determination whose effects cannot be witnessed, and the uniformity of the mechanical laws of development that preside over the deployment of organisms constituted from pre-existing germs. With the combination of subsequent effects, both terms unite, i.e., organized living bodies and their functional properties, obeying divine power and wisdom, exerting themselves within the innate generation of beings by the particular constitution of germs and regulating their deployment or "unfoldment" via sequences of transformation that can and must be presumed to be true and subordinate to the laws of constituted nature.

The conception of these developmental laws, proposed by Vallisneri, a priori reducible to mechanical models and to modes of analysis unique to the science of natural phenomena, requires more in-depth investigation. Roger presumes that this question remained unexamined in the analyses produced by the author of *Istoria della generazione*[4] and that it is with Louis Bourguet, in his *Lettres philosophiques*, that the question was broached in a new manner. Certainly, the organizational activity of nature cannot be conceived of without postulating pre-existing germs. That is, apparently, a necessary postulation, but one must also account for the mechanisms active within development, which is a major task of the science of living beings,

[3]Quoted in Roger (1971, 374)

[4]"From a strictly scientific perspective, Vallisneri brought, ultimately, nothing more than a very lucid clarification, and what is interesting in his intervention resides mainly in his general reflections on the knowledge of nature, to which he found himself driven. In particular, he did not pay special attention to the question of the development of the pre-existing germ" (Roger 1971, 375–376).

prone in fact to act retroactively on the very comprehension of the postulate of the pre-existence of germs.

One of Dario Generali's major contributions to Vallisnerian studies consisted in emphasizing the light shed by unpublished sources on published works. This is particularly true of the interpretation of *Istoria della generazione* to be provided. Yet, some aspects of the genesis of this work can be explained by the triangular correspondence between Leibniz, Bourguet, and Vallisneri and by Leibniz's recommendation to Vallisneri, with Bourguet as intermediary, to devote focused analysis to the foundations of rival theories of generation, so as to decide among them. The nature and scope of this exchange has been the subject of a study from which I will, to satisfy my purposes, only retain the basic points.[5] Therein, the Leibnizian incitement, made to one of the most well-informed naturalists, to choose among the explicative hypotheses of generation, and more particularly the ovist and animalculist preformation models is vividly related. Indeed, Leibniz considers that Leeuwenhoek's theses show a priori as much interest and theoretical validity as those of the opposing party. Taking for granted the basic principle of pre-existence of sufficient organic structures, he thus asks for empirical criteria, tied to observation and experiments, which might likely ensure the prevalent corroboration of one of the theses.[6] It's to take up this challenge, while refuting Leeuwenhoek's animalculism, that he urges Vallisneri with a series of arguments. As established by Generali, Vallisneri, who has read *Théodicée*, manifestly feels an affinity with Leibniz's positions on the system of nature, on the chain of beings, and on the status of organic bodies as natural machines: stemming from creation's primeval plan, these are believed to pre-exist, in enveloped organic form, to the series of successive generations. He manifestly also tried, with *Istoria della generazione*, to answer the question of the differential validation of opposing hypotheses that Leibniz had formulated, as witnessed by the letter that Vallisneri addressed to Bourguet on 13 February 1717, soon after Leibniz's death (14 November 1716): "You, Sir, would

[5]See Generali (1987, 125–140).

[6]See the letter, Leibniz to Bourguet, 3 January 1714, GP III, 562: "Mr Vallisnieri is very solid. He contradicts Mr. Leeuwenhoek on seminal animals; but I would want the fact to be made clearer, regarding eggs as much as these animals. I am nevertheless certain that an organic body is never formed from chaos or from an inorganic body, and even that there is chaos only in appearance." See also the letter, Leibniz to Bourguet, 22 March 1714, GP III, 565: "[...] were that to be true, the question would always remain, if the basis of transformation or preformed life is in the ovary, as per Mr Vallisnieri, or in the sperm, as per Leeuwenhoek. Because I maintain that there must always be preformed life, be it plant or animal, as the basis of transformation, and including the same dominant monad: no-one is more apt to elucidate this doubt than Mr Vallisnieri, and I wish greatly to soon see his dissertation; to be the object of its dedication would honour me more than I deserve." See, finally, the letter, Leibniz to Bourguet, 11 July 1714, GP III, 571: "I would be most curious to one day learn to what Mr Vallisnieri objected to in Mr Leeuwenhoek, and I would strongly wish that this should appear while Mr Leeuwenhoek is still alive [...] I do not know whether these animals that find themselves in the semen of larger animals perform the function attributed to them by Mr Leeuwenhoek, but I do believe that, if this function is not performed by these animals, there are other, invisible ones who perform something similar, since it appears that we cannot avoid a pre-existing animal."

do very well in publishing the Leibnizian philosophical system, since there are most beautiful things in it [...]. I have already done much work on the subject of spermatic worms and, though Mr Leibniz is dead, I shall publish the *Istoria*."[7]

But, beyond some general affinities, it's not as obvious that Vallisneri grasped the nature and scope of Leibnizian theories, such as those on power and on composite substances, let alone their effect on the interpretation of the notions of monad and of organic body, which emerge from final presentations of the system, such as *Principes de la nature et de la grâce* and *Principes de la philosophie ou Monadologie* (1714). In fact, the most profound and authentic Leibnizian position deals with the conception of a properly organic integration of the material structures of the living being at various stages of development. This integration would be conditioned by the states of the dominant monad in its interrelations with subordinate monads, which would represent the formal reasons for the diverse combinations of material parts temporarily integrated into organic structures. Moreover, the central determination of this model consists in the autonomous power of action passed on to the organism. The organism produces its own transformations, obeying an internal law of development; such a development is certainly predisposed, but it produces strictly architectonic effects from the original organic base. However, from a methodological perspective, such effects can and must be explained in a correlative and complete fashion by resorting to complex mechanical models that provide their analytical equivalent.

It's that dynamic side to the Leibnizian doctrine of organism, apparently not as present in Vallisneri, that Bourguet attempts to clarify in his *Lettres philosophiques*, while also being inspired by a theoretical framework analogous to the one being developed by the author of *Istoria della generazione*.

As for the theory of generation, Bourguet's position is original for the period when he formulated it. It is a foremost theory because it is based on an attempt to give a unitary explanation of organized bodies by combining that which, in the organization of life, is attributable to laws of a physical sort—relative, on the one hand, to the effect of the forces at work and, on the other, to the processes of crystal production—with that which, in contrast, by reason of its irreducible characteristics, would no doubt cause a breach of intelligibility, unless we allow for an original organic production that would constitute its sufficient reason. Bourguet is original because he pushed as far as possible research into a reductionist explanation of a mechanistic type, so as to combine in the end that approach with a postulate of original organic entities—prior and pre-existing organic structures susceptible not only to simple development but to real transformations determined by the interaction of closely related causes of a physical-mechanical type. Pre-existing germs are thus believed to be not so much complete miniature organisms but rather organized matrices whose organization and processes are regularly completed by an "organic mechanism" exerting itself in the integrative structures enveloping germs (Bourguet 1729, xxxviii and passim). As highlighted in

[7] Bibliothèque de Neuchâtel, ms. 1282, cc. 219–220, quoted in Generali (1987, 130 n. 29).

the preface to *Lettres philosophiques*, the germs of plants and animals are organized bodies, and we could not assume that they are formed by "the combination of non-organic particles," let alone by "a mechanical collection of already organized particles" (Bourguet 1729, xxxvi)—this second mode being that of the composition of crystalline structures by the joining of "molecules of the same figure." Under these conditions, it is necessary to refer to a principle of divine "predelineation"—an expression that Bourguet links to Leibniz's pre-established harmony.

In the same breath, however, Bourguet objects to the thesis of the simple growth of preformed organic structures. Because of germs' weight increase, but mostly because of the observable structuring of various tissues during embryonic development, he maintains that the liquids that operate in the expansion of primordial structures must undergo a prior transformation, and then, an integration into already present organic matter, which must be duplicated according to specific compositional characteristics and structural dispositions. From that perspective, Bourguet places the thesis of the organization of material corpuscles and, more specifically, the organization of infinitesimal plant and animal embryos—embryos that possess "active principles" (Bourguet 1729, xl) comparable to those of monads—under the aegis of Leibniz. Such is the order, on the one hand, of the primitive constitution of organized bodies and, on the other, of the immaterial substances that underpin them. The evolution of bodies and monads must answer to immanent laws that regulate the natural order. Yet, the laws that preside over the transformation of organized bodies properly define the "organic mechanism" that exerts itself on and via primordial organic structures.[8]

Vallisneri is one of the honorific addressees of *Lettres philosophiques*, and how Bourguet's theories relate to Vallisneri's arguments is one of the key elements of the argument of the third letter, while it also sketches a critical evaluation vis-à-vis the Leibnizian theory of organic bodies. The third letter deals with spermatic worms, the dust of flower stamens, embryonic molds, plastic natures, and guiding minds. The initial theme is the nature and function of spermatozoa, which were discovered by Nicolas Hartsoeker (1656–1725) and Antoni van Leeuwenhoek (1632–1723) and identified by some, such as Nicolas Andry de Boisregard (1658–1742) (Boisregard 1700) as the pre-existing and encased primordial germs of all animal organisms. Bourguet situates Leibniz among them, but this assertion is somewhat forced, as near the end of Leibniz's career, in a correspondence between the two men, Leibniz, without excluding that hypothesis, places it on the same footing as the hypothesis of ovist preformation and avoids categorically endorsing one over the other.[9] Bourguet's rejection of the animalculist hypothesis

[8]This approach is clearly present in the letter of Leibniz to Bourguet, 5 August 1715, GP III, 579: "I could not say anything about the details of animal generation. All that I believe to be able to assert is that the soul of every animal is pre-existent and has been in an organic body that, ultimately, by way of many changes, involutions, and evolutions, has become the current animal."

[9]See the letter, Leibniz to Bourguet, 3 January 1714, GP III, 562, quoted in Duchesneau (1998, 328).

is assigned to the critical reflections he elaborated in Venice in 1714 and that he was delighted to see integrated in the *Istoria della generazione*: "Mr Vallisneri's excellent treatise on the generation of Man and of the animals, printed in Venice in 1721, not only includes all my ideas, for which I congratulate myself, but also includes many others that, to my mind, make the matter incontestable" (Bourguet 1729, 76).

We must understand that Bourguet does not find all the elements of his theory of generation in Vallisneri but only the elements specific to his critique of the animalculist hypothesis; Vallisneri's treatise completes these elements with additional specific arguments. More than any other argument, it is the generation of relatively complex animal organisms without the participation of the presumed male seed, for example in the parthenogenesis of aphids, that underpins his critical position on the subject of encased "spermatic worms." At the same time, the spermatic worms are considered as true animals possessing their own ovist encasement category, as they could not be considered the product of spontaneous generation; rather they were thought of as a sort of parasitical organism observable in the semen of males that could also be imagined to exist even in females, who would be responsible for transmitting them to their offspring. The explicit objections against the role of the presumed germs, which Andry attributed to spermatozoa, related to (1) the prodigious disparity of the number of spermatozoa versus the number of foetuses; (2) the extreme difficulty to conceive of, according to the mechanics specific to organic bodies, a model able to explain the selective penetration of a worm in one egg, to the exclusion of all others; (3) the small size of the worm relative to the size of the egg that would serve as its envelope (confused with Graaf's follicle)—the growth of the resulting embryo would seem to be too rapid and out of proportion "following the usual laws of movement by which grow all organized bodies subject to this order" (Bourguet 1729, 86); (4) the quasi-impossibility of conceiving that the parts of the egg that constitute the foetus be the integral parts of the worm and not only parts linked to the worm. On that subject, Bourguet points to the arguments developed by Vallisneri on the imperceptible—because it is transparent—but relatively complex and integrated structure of the foetus as it would pre-exist in the egg prior to embryonic development; this structure would be better proportioned in relation to the embryo that emerges from it than any structure inherent to the spermatozoon.[10] As to the role of male semen and, correlatively, of stamen dust in plants, it seems to be reduced to a function of densification and activation of liquids that act on the germs and seeds so as to stimulate the movements required for organic development.

[10]Cf. Bourguet (1729, 88): "And even if Mr Vallisneri admits with much frankness that he has never seen the eggs of viviparous animals in their cells, I suspect that the yellow or glandulous body is the egg itself covered in a film and that the triangle that this learned man has found in its centre corresponds to the one we see in the eggs of oviparous animals. If this idea of the embryo supported by the analogy of the nucleus and the seed occurs, as I believe all experts will admit, the retortion could not subsist as the proportions in the embryo's growth are better retained in our system than in the one we reject."

6.2 The Nature of Organized Bodies

In the introduction to *Lettres philosophiques*, Bourguet already takes a stand in favor of a conception of organized bodies that rests on Leibniz's theses regarding organisms, which are presented in a series of post-1704 Leibnizian texts and which extend the revision of the system of nature following the invention of dynamics per se in 1689–1690.[11] I presented the Leibnizian theory of organic bodies in, among others, *Les modèles du vivant de Descartes à Leibniz* (Duchesneau 1998, 315–372). Here, I will merely highlight the elements retained by Bourguet, who does not fail to bend or alter them somewhat to link them, on the one hand, to the empirical issues that solicit his attention and, on the other, to the theoretical divergences that are emphasized in the immediate context of *Lettres philosophiques*.

Bourguet adopts a concept of the ultimate elements of the phenomenal world that directly links to his theory of force and action inherited from Leibniz. Thus, he asserts that all the "corpuscles of matter," including those of the least degree, conceal an "organization" that links them to "active principles" comparable to Leibnizian monads (Bourguet 1729, xxxix–xl). Within the same pages is reiterated the serial and hierarchical gradation of organisms, from that of the smallest corpuscles of matter to that of the "infinitesimal embryos of plants and animals," from the latter to that of the complex living organisms that emerge from them after transformations due to "organic mechanism" and, by analogical extension, to that of realized human organisms, which comprise highly integrated structures, underpinning the physiological processes correlative to specifically intellective and volitive mental activity. From the start, understanding this system presumes principles of harmony that link the mechanisms at work in these various stages to developmental laws immanent to corresponding immaterial substances.

Two major traits characterize Bourguet's importation of the Leibnizian theory of organic bodies: a concept of material organization as expression of the inherent limits of the active principles at work; a concept of a chain of beings as an integrated expression of infinitely diverse limits specific to nature's various living machines. Regarding the former, we must conceive of immaterial substances—the Leibnizian monads, which are essentially active and alive—as strictly correlated to determined portions of organized matter. Bourguet does not seem to appropriate the Leibnizian theory of composite substances as integrated products of organized bodies and of their respective monadic principles, the organized bodies themselves integrating the effects attributable to subordinate monads by virtue of the degree of activity of the dominant monad that constitutes them, but his conception does not stray far from it and does not substantially differentiate itself. Thus, he asserts:

> Everyone agrees that creatures necessarily have limits and that these limits can only be the reciprocal resistance of their activity. Because, if they didn't reciprocally limit each other, their activity would extend infinitely, as nothing would stop them. Thus, they would simultaneously be both creatures and non-creatures, which is an obvious contradiction. These

[11] On the invention of dynamics, see Duchesneau (1994, III–IV).

> limits are nothing more than what is passive within created beings, whose actions and reactions complete activity within a regulated subordination following the perfection of each in particular. These limits are thus nothing more, really, than what we call matter, because the details of the united action of the agents that resist us, or that complete our action with their own, are imperceptible. We conceive of it as a whole under an enveloped idea, which is why we call it matter or body—i.e., a multitude that can only be discerned as a whole. (Bourguet 1729, 112–13)

The limits inherent to the realities of the natural order are of two types. The first type corresponds to the limits specific to the activity of each monad and defines the specific "circumference" of the ensuing functions, i.e., the limits of the actions that the monad is able to actualize in the phenomenal world by the organization of its own body as a vital machine composed of determined organs. The second type of limit comes from the correlation of the individual spheres of activity among each other; these spheres seem to interact as a function of the dynamics of their particular respective bodies, the ones acting, the others reacting in a reciprocal fashion and in accordance with the action potential enveloped within the very composition and structure of these organic bodies.

From the same perspective, Bourguet appropriates the thesis of the great chain of beings in its Leibnizian version.[12] Finite substances are thus conceived of as forming a gradation of perfections by virtue of their intrinsic activity, reflected by the series of organized bodies. In turn, this series comprises an "infinitely regular whole"—the world, the universe—that is nothing but "the result of the phenomena of activity of all these beings combined" (Bourguet 1729, 115). Each organism constitutes, in its way, a summary of the world, by reason of its infinite actions and reactions of which the organism is both subject and object, but, according to the degree of complexity of the organic machine and the degree of refinement to which it is susceptible in its action, the organism in question finds itself within a hierarchical series that, it must be presumed, obeys the architectonical principle of continuity. We should note here that the empirical understanding of this series of gradual perfection in organization cannot be reduced to the gradation of a single trait but can be to the gradation of the internal laws of composition of structures and integrated deployment of functions, according to what we could consider as a principle of matrix order underpinning the empirical analysis of phenomena. However, the principle of continuity is used here to justify the ordered maximal proliferation of organic entities, by virtue of the infinitely varied omnipresence of lifeforms and of corresponding organized bodies. Concerning the presumed status of spermatozoa as microscopic parasitical organisms, Bourguet refers in fact to the following version of the principle of sufficient reason; he writes:

> It's a matter of knowing that there is no part of matter where there are no animated organic bodies; and as there always remains matter that is not necessary to this or that animal, the supreme wisdom did not want this matter to be entirely useless and made it of use to other, smaller to the infinite, animals. (Bourguet 1929, 92)

[12]Concerning the Leibnizian interpretation of the great chain of beings, see Duchesneau (1993, 359–74; and 1995, 47–59).

At the end of the fourth letter, to support the Leibnizian model that he adopts, Bourguet uses the very arguments that he gleans from the analysis of generation and the critique of various explicative reasons. He then continues with a specific configuration of the distinctive traits of that model, based on the version he provides. Let's assume that organized bodies possess a structure that can integrate in a sequence of continuous changes corpuscles of an inferior degree without substantial change to the structure of these more elementary composites and no change either to the economy of the organized body of a superior level. At the same time, the processes of organic mechanism obey the determinations of the immanent law of assimilation-disassimilation of corpuscular parts, a characteristic law of the type of organism. "The molecules that enter into organized bodies," infers Bourguet,

> can, by uniting and separating themselves, form all necessary changes without there being any real transformation to the interior of things. By staying as they are, they meet all their requirements, thanks to the mechanism that God has instituted from the beginning. The very bodies of plants and animals are tiny worlds, infinite series in their way, that contain an infinity of other series in expressions that are lesser to the infinite. (Bourguet 1729, 165–66)

Yet the status of organic body as microcosm implies, at the lowest degree, correspondence to a "monad," a "principle of life," as per Leibniz's terminology, or to an "active principle," as per Bourguet's terminology, consisting in a "continual tendency to act" (1729, 166), which must be conceived of in accordance with the analogy of a force realizing its effect in action by a summation of impetus. The evoked analogy is that of a tensed bow that is released, or that of a coil that springs. At the inferior degree, Bourguet asserts, the active principles are not "accompanied by emotion and perception": here, we must understand that they do not imply "apperception" in the Leibnizian sense, but, in the same breath, we should note that Bourguet pulls the Leibnizian monad—endowed with perception and appetition—in the direction of a monad that is a simple immaterial (or infra-material) subject of power and action. The expression of the plain monad in the physical order does not consist only in modalities of figure and movement delimiting the sphere of action and reaction that is attributable to it. The progression of activity and of structural perfection that the monads envelop leads, by degree, to the souls of plants and animals, by way of transitions that symbolize, respectively, sensitive plants and zoophytes. The analogical sequence then leads to the recognition of the status of the soul of man, which is heightened by intellection and will.[13]

But every monad, whatever its degree of perfection, is more or less passive, as it is more or less subject to "the general mechanism of bodies," which characterizes the order of causal sequences in the physical universe. In a specific sphere, that of vegetable and, especially, animal beings, that mechanism takes the shape of a specific mechanism: "organic mechanism." It must certainly not be understood here that the determinations of a physical order are more or less suspended based on the power to produce contingent effects possessed by monadic agents. Without a doubt,

[13] According to Bourguet's formula (Duchesneau 1993, 167): "[...] reason and freedom in the choice of the objects within its reach."

the laws of mechanics and of the physical order can be applied integrally to the entire universe of bodies, including organic bodies; at least, that is the strict lesson that can be gleaned from the scientific method and from the Leibnizian system of nature. But, at the same time, Leibniz assumed an integral correspondence of the order of purposes to that of efficient mechanical causes, and a relative autonomy of functionality possible for the more complex organic subjects following the infinitely complex integration of the mechanisms that constitute them according to a law of harmony. That's what Bourguet explains with illuminating formulas, but that are equivocal because they omit the analytical understanding of contingent truths, without which, according to Leibniz, we could not unlock the epistemological paradox in question. Indeed, Bourguet is satisfied with asserting:

> Though this passivity submits [monads] more or less to the general mechanism of bodies, and to organic mechanism with regards to a portion of them, that does not prevent the rules of movement from accommodating themselves to the spontaneity of some and the freedom of others, and spontaneity and freedom from, in turn, being in tune with the rules that the divine wisdom has found appropriate to establish when it gave them existence or when it created the world, which is, in my opinion, completely the same thing. (1729, 168)

6.3 The Role of Organic Mechanism in the Explanation of Generation

Following the methodological approach that he favors and that he undeniably shares with Vallisneri, Bourguet assigns to the desired explicative hypothesis of generation the obligation to account for, in as satisfying a way as possible, all known phenomena. This hypothesis must furnish the full and sufficient reason for the empirical observations that are accepted or that are considered as highly probable in regards to the processes of generation.

These processes are empirically characterized by the generic effects noted below. These constitute the explanandum that the theory must account for:

> *The Phenomena*
>
> In the union of the sexes, males and females each give a liquid; it is to the mixing of these that we attribute the conception or fertilizing of the egg, when they reach the ovary via the Fallopian tubes, which then usually approaches it.
>
> The foetus has several organic parts so absolutely necessary while it is in the mother's womb that it could not subsist a single moment were it deprived of them. That phenomenon is generally observable, albeit with some diversity, in all viviparous and oviparous animals, not excluding insects. It is observable also in the seeds of plants.
>
> The small animal in the egg and the small plant in the seed are folded into a point beyond our imagination and that becomes perceivable shortly after its fertilization. We then observe that their organs are so interwoven among each other that they occupy the smallest possible space.
>
> The foetus usually resembles the male and female that engendered it. We observe in all families that the children take on sometimes more and sometimes fewer of the complexions

> and traits of the father and mother or of the people of the family of the one or the other. The foetuses also usually inherit their infirmities.
>
> Animals that come from the conjunction of two different species partake also of the nature of both.
>
> The foetuses of animals that have a vivid imagination and upon which objects act with force often show visible marks that manifest themselves in brutes in the diversity of colors and in children in the face, and sometimes in the color.
>
> From time to time, monsters are born, which occurs in plants, in animals, and in Man, with this difference: it happens more rarely in the first and more often in the last two species of organized bodies. Monsters can be reduced to three categories: (1) the deformed; (2) the defective who lack some essential part; and (3) those who have too many parts. We could also add the offspring of two different species of animals. (Bourguet 1729, 93–95).

The explanans takes the form of the various hypotheses that Bourguet brings back to three principals: the formation of organisms by molds, their formation by plastic natures, and finally their initial dependence on a transcendent architectonic cause that enables the mechanical derivation of all subsequent effects.

The mold hypothesis includes the notable premise that "the organic produces the organic." Without there being any precise reference to some prior doctrines, we must grasp here the expression of mechanical epigenetist forms that presume a specificity of produced bodies in accordance with the organization and the internal composition of the originating organisms. Such ideas were, for example, expressed in the first proponents of a corpuscular structure of organic matter, a structure modeled after the structural and functional combination represented in the parent.[14] This is undoubtedly aimed at the Gassendist tradition, rather than the Cartesian tradition or that of the Aristotelian type of epigenesis, which was still represented by William Harvey.[15] Here, the central idea is that of the fashioning by an organic body of a similar organic body, by virtue of the constancy of a matrix structure exerting its organogenetic power on "organic molecules." This matrix structure—or "mold"—is purely material or implies organizing agents outside the mechanical order.[16] When, in this context, Bourguet speaks of "idealists" whose opinion harkens back to the theory of molds, he is undoubtedly alluding to the followers of doctrines of the Van Helmont type. The function of reference for the epigenetist corpuscularist group is nutrition that is based on the transformation by digestion of the molecules absorbed into chyle and on their transportation via the blood flow toward the organic parts, which they augment by assimilating themselves into them. Yet, the analysis of this process suggests two characteristics of the organisms in question: on the one hand, the eminently complex organization, both and simultaneously differentiated and

[14]On the Gassendist theory of generation, see, among others, Duchesneau (1998, 85–118).

[15]See French (1994).

[16]It has been emphasized several times that the concepts used here by Bourguet will have a significant new life in a neo-epigenetist context, notably in the theory of generation expounded by Buffon in the second volume (1749) of *L'histoire naturelle générale et particulière*. See, for example, Roger (1971, 378 and 546).

integrated, of the parts that need to be developed by a form of reproduction; on the other hand, by virtue of the dominant orientation of the mechanistic explanation in the moderns, the presupposition that this diversified and harmonious effect is realized "by a mechanism that, being the same in all [organs], is particularized in each" (Bourguet 1729, 96). Ultimately, the mold in question can only be "the mold of the foetus's entire body" (Bourguet 1729, 96).

Yet, that mold would be located in the egg itself or in the neighboring parts of the parent organism, or would correspond to the entire body of that organism. This last option, which agrees with the ancient tradition of defluxus, seems at first glance the most admissible. There remains, of course, to explain, in the context of bisexual reproduction, how fragmentary roughs, corresponding to the disseminated organic parts of the parent organisms, find themselves harmoniously joined to form a unique, integrated, and functional organism. Hence the suggestion of an additional and specific mold able to accomplish this complex organogenesis. To represent this mold and its properties, the physiologist tends to resort to a particular animistic principle: "the soul of brutes that resides in sperm and that forms their bodies by a plastic virtue, and by something analogous in Man that, under the direction of reason, itself builds its domicile" (Bourguet 1729, 98). The mold theory inescapably leads us from a corpuscularist and mechanistic epigenetist conception of the formation of organized bodies to a more adequate and satisfactory reason devised in the form of plastic natures. From the beginning, a purely mechanical formation could not, in fact, account for the subtle and functional arrangement of the parts that the embryo itself reveals. If the idea of a mold shouldn't be repudiated, it's on the condition of a pre-existing organic formation and of a development—a transformation, even—of this innate structure by the interplay of the mechanical forces of the natural order regulating the contribution of new matter to the minimal organism.

In the particular historical context during which Bourguet deploys his analysis, the recourse to plastic natures represents a major option for bypassing the lacunae that affect mechanism when one tries to explain the primordial and integral formation of organisms. The importance and strength of this trend should not be underestimated. I have retraced its roots in the Cambridge neoplatonists Henry More and Ralph Cudworth (Duchesneau 1998, 149–182). I have also considered the updated forms, from the early eighteenth century, in the presentation offered by Jean Le Clerc, which ignited the controversy between Le Clerc and Pierre Bayle (1647–1706), with Leibniz's subsequent intercession (Duchesneau 1998, 346 n.1).[17] But it must be noted that the trend of plastic natures annexes multiple other

[17] To counter the mechanism applied to the conception of living beings, Le Clerc had taken on the task of exhuming Ralph Cudworth's major work, *The True Intellectual System of the Universe* (1678), and of publishing synopses of it in *Bibliothèque choisie* (1703–1706). Following the same line of thought, he supplied excerpts from naturalist Nehemiah Grew's *Cosmologia sacra* (1701). Suspecting that atheism would benefit from the theory according to which spiritual plastic natures would form and animate the living in an autonomous fashion, Pierre Bayle had jumped into the fray as of August 1704, by developing a polemical exchange with Le Clerc (1704–1706). Leibniz found

theoretical variants, beginning with the animist doctrines of Claude Perrault (1613–1688) and Georg Ernst Stahl (1660–1734) (Duchesneau 1998, 265–314). From the perspective of *Lettres philosophiques*, Hartsoeker's critical essays regarding animalculist preformationism (which he had previously maintained) relate to an analogous tendency. It is also curiously obvious that Newton's empirical methodology, as he presents it in the "Queries" of the second English edition of *Opticks* (1717–1718), and the analogical inference allowed by this presentation facilitate the conception of specific gravitational forces: wouldn't these be prone to underpin complex organogeneses in a world for which resolution into mechanistic models could not be integral, as human understanding would suffer from deficient access to the deep meta-empirical causes of basic physical phenomena?

Bourguet associates the recourse to plastic natures to an acknowledgment of the intelligibility deficit that any mechanistic model would represent in relation to the formation of a machine of nature as complex and as delicately arranged as any living organized body. The mechanistic explanation could only lead us to take into account the productions of crystalline formation type, which, indeed, are regular but are distinctly below diversification and, simultaneously, below the integration of structures and of functions inherent to living organized bodies. Must we, under these conditions, assume an immanent intelligence in the natural order that performs this demiurgic function? The analogy with our own capacity of intellection dissuades us from entertaining such an hypothesis, as our own technological power reveals itself to be incommensurably deficient when it tries to understand the innate production of the least plant or animal.

> It is not by an ingenuity similar to Man's that these beings can form organized bodies; because, as the bodies of plants and animals are machines whose excellence infinitely surpasses all that the capacity and ingenuity of mankind can produce, it would be necessary, to form the least of organized bodies, to possess knowledge and ability infinitely superior to mankind's, and instruments of infinite sensitivity to execute this. (Bourguet 1729, 117)

Of course, the proponents of plastic natures willingly, in lieu of a rational intellection immanent to organisms, propose a form of architectonic instinct that acts in accordance with the analogy of behaviors that are infra-rational but also producers of order and vital organization in numerous animals. The counter-argument proposed here by Bourguet rests partly on the respect of a fundamental principle popular among physiologists, notably those of the Malpighian tradition: the principle of strict correlation between microstructures and the functions deriving from these. This counterargument also rests on the principle, clearly laid out by Leibniz, of a rigorous correspondence between the functional capacities of the living machine

himself implicated by Bayle, the defender of occasionalism against all possible restoration of substantial forms. Furthermore, Le Clerc appealed to him to supply an opinion on the opposing theses and to clarify his own system. As a result, Leibniz decided to distance his theory of the organism from the revalorized doctrine of plastic natures: thus, in 1705, he published "Considérations sur les principes de vie et sur les natures plastiques" in *Histoire des ouvrages des savants*, edited by Basnage de Beauval.

and the internal representative states of the dominant monad in relation to the subordinate monads concerned with the various parts of the machine of nature. The architectonic powers that can be integrated into the properties of the most perfect monads of the natural order could not figure among the properties of the least perfect monads. Yet, the powers of our souls do not include the power of producing the complex organizations that those very souls depend on for the expression of their characteristic actions; more conclusively, it is even the more so in regards to the production of animal or vegetable bodies of an inferior order, whose architectonical plan surpasses any finite capacity of intellection and volition.

Of course, against this deployment of the double principle of correspondence evoked by Bourguet—a correspondence between structure and function as defined by the micromechanists, and a correspondence between the functions of the perceiving agent and the dispositions embodied in the composition of the body itself—Hartsoeker had already put forward his interpretation of Réaumur's experiments on the regeneration of amputated pincers in crawfish.[18] The claws's mechanical formation is confirmed, which supposes an intelligence immanent to the organic device: should we not assume that the very formation of the animal is attributable to the same type of process, and thus to a manufacturing intelligence of this immanent type? And so it would be in regards to the production of spermatozoa by an immanent intelligence operating in the male sexual organs. But this intelligence inherent to the organism and able to organize it, regulate it, and move it in its functional activities could not be, in this view, distinguished from an organizing power disseminated throughout the entire organism—in the entire field of natural phenomena, even—for all that these phenomena express a real organization and dynamism, just like the crystalline formations that Bourguet distinguished from true organized living bodies. Hartsoeker assimilates the correlated terms of the principles of correspondence; Bourguet, who is faithful in this to the argumentative logic of the micromechanistic naturalists and of Leibniz, critically differentiates the same terms. Indeed, Roger quotes a passage from *Recueil de plusieurs pièces de physique* (1722) that excellently evokes the analogous assimilation process at work in Hartsoeker, who, on this point, is the heir of the plastic nature and physiological soul theorists. In fact, Hartsoeker, a proponent of hylozoism, writes:

> Ultimately, I opted for conjecturing that there is within us one soul only that does everything inside of us. Furthermore, I conjectured that this soul could not be anything but a portion of the soul of the Universe, which, in my writings on physics, I have called the first element or the perfectly fluid substance and which, being spread like matter, although in all other aspects it is essentially different from it, can drive bodies and be driven by them, give them movement and receive it; I was even more determined to make this conjecture because, with it, I freed myself from the great difficulty that we have always had of conceiving how soul and body communicate with each other. Thus, I conjectured that the soul is spread throughout the body. (Harsoeker 1730, II, 68)[19]

---

[18]See Hartsoeker's letter to Le Clerc in *Bibliothèque ancienne et moderne*, VIII (1717), 303–350.

[19]Quoted in Roger (1971, 434). The *Recueil* was republished in the *Cours de physique* (Hartsoeker 1730).

In reply to Hartsoeker, Bourguet concretely puts forward the eminently limited character of the reconstitution of the crawfish's pincers: it is a reconstruction strictly normed within the framework of an organism that is already built so as to allow for the realization of such a phenomenon. To this end, regeneration occurs in specific locations, in accordance with a process whose unfurling is in some way programmed and which is akin to the simple deployment of a pre-existing structure. But, especially, the process reiterates itself only a limited number of times when we amputate the reproduced pincer, which seems to confirm that "this reproduction is proportionate to the natural or probable peril of the loss that [the crawfish] can expect and to the ordinary lifespan of the animal" (Bourguet 1729, 126). This argument is remarkable: it points to a programming of these successive regenerations that is in some way mechanical, within the context of a predetermined type of life and of a predetermination of the structures required to meet, in accordance with the natural order, the external accidents putting the integral organism in peril. What seems to be preprogrammed is in fact a function of safeguarding that integrity, embodied in the structural devices of the type of organization in question. Indeed, Bourguet remarks, in regards to the regeneration of crawfish pincers, that the processes at work represent a form of vegetable production within an animal structure and that, accordingly, they would be even more obviously linkable to a regular and mechanical order of deployment of the initial organic device; notably, he mentions Vallisneri's explicit agreement on this point.[20] We can infer that the phenomenal characteristics of regeneration processes prevent their attribution to plastic natures or to physiological souls; on the contrary, they would push us to link them to an "organic mechanism" able to provide the most appropriate analytical grid for all phenomena needing to be explained.

In the fourth letter—"in which is explained the system of developments and the organic mechanism" (Bourguet 1729, 132)—Bourguet proposes a general thesis in accordance with the epistemological model maintained by, among others, Malebranche; according to this model, the conception of the encasements required by the theory of the developments of encased structures could not be assessed by the yardstick of simple imaginative representations but in accordance with a conceptual evaluation of the explanation's requisites. A substantial part of the argument thus developed consists in challenging Hartsoeker's calculations that seek to limit, to their complete exclusion, the number of encased germs; the reasons that are brought up deal with the disparity of the extensive dimensions of these germs in serial order, if we only presume a distant temporal origin for pre-existing organisms and their germs. We will not follow the details of the reply, which is inspired in large part by considerations and calculations developed by Vallisneri. The emphasis, it must be noted, is particularly on the dimensional transformations of the germs themselves, which are no longer considered simple miniatures of the completed organism but

[20]"[. . .] the crawfish, which has many parts that renew themselves from time to time, via a development very much akin to that of plants (Mr Vallisneri, one of the most precise and skilful observers of nature, is of a similar mind regarding crawfish, which he took the trouble to inform me in a letter of the month of December 1723)" (Bourguet 1729, 148).

as the most minimal structures potentially comprised of the characteristics of the emergent organization.

In correlation, as Bourguet puts forward against Jean Le Clerc (Bourguet 1729, 140–41), the presence of monstrosities and anomalies in the products of generation does not contravene the thesis of a primeval formation of organic devices, as the development of these devices falls within a system of nature where dominates a contingent application, because it is circumstantial, of the laws regulating that development, including the laws of organic mechanism that regulate more specifically the development of preformed microstructures.

Admittedly, the conception of "organic mechanism" is constructed around a premise purporting that original vital organizations could not be explained by any rule of mechanical production. These primordial organizations are nature's machines, irreducible to the machines of the art as to the sufficient reasons of their very existence. Their arrangement is so complex and integrated that we could not deduce their origin from the application of any mechanical law to organizations of an order of infinitely lesser complexity. Of course, we can compare organized bodies to clocks with different structures—even to looms of multiple and diverse registers (Bourguet 1729, 143–44)—but the analogy cannot hold when we consider the application of rules of mechanical operation to hyper-complex structures that already exist; correlatively, if these are from the start subject to an infinite reduction of their spatial dimension following the order of calculation, we could only conceive of their production under a mode of a totality produced in the moment. Bourguet asserts: "[...] Plants and animals are thus like infinitesimals; [...] Thus, the almost infinite smallness of their volume and the arrangement of their organized parts necessarily presuppose an instantaneous operation; i.e., to produce them, an action that in one swoop made the organs and their arrangement was necessary" (Bourguet 1729, 143).[21] It follows that the organic mechanism can be applied to pre-existing organisms duly constituted; but it can also be applied indubitably to their sequences of transformation and thus to the emergence of embryonic organisms that result by development. Bourguet asserts that:

> [. . .] Organic mechanism could only happen in an already organized body and [...] it's a fundamental axiom in this matter that we must necessarily allow for pre-existing organized bodies, before organic mechanism can act. After that, it isn't difficult to imagine that the same mechanism that had acted in the larger body might be communicated to the smaller one that it contains and could, in that smaller body, produce effects identical to those that it had produced in the body that preceded the smaller one in order. (Bourguet 1729, 146–57)

The processes of development, nutrition, growth, and spontaneous movement are related to organic mechanism: these processes represent the sum of operations ascribable to living plants and, to an even more complete extent, the sum of operations ascribable to animals. These operations occur in different ways according to the various modes of reproduction of organisms, whether that reproduction depends

[21] Note, in this passage, the metaphorical use of the notion of "infinitesimal," inspired by Leibnizian infinitesimal calculus.

on the participation of both sexes, or the formation is endogenous or it is parthenogenetic. Development occurs when the parts of the preformed organic body are subject to a characteristic movement under the effect of the movement inherent to the seed or to the mix of seeds. The seed is seen as "an extract of the parts of the animal that communicates it" (Bourguet 1729, 149). In cases of copulation, the mix of spirituous liquids extracted from the parent organisms acts upon the parts of the embryonic micro-machine, inserting itself within it and supplying it with the initial nourishment needed to jumpstart its growth and the deployment of its structures and of its emerging functions. In the Fallopian tubes, then in the matrix, nutrition operates according to specific modalities, until the organism reaches the state of autonomous life. The development of the embryo follows a sequence linked to the differential and progressive assimilation of the molecules most likely to develop its respective parts: bones, flesh, nerves, membranes, etc., which become visible one after the other because of their growth. Nevertheless, as basis for these organogenetic processes, a prior integral formation of the embryonic organism is assumed. The illustration of this point is tied to Malpighi's paradigmatic observations on the formation of the chicken in the egg.[22] The crucial point of the argumentation deals with the refinement of the primitive organism, which is able to execute the replacement and differential assimilation of molecules, at the same time that the liquids acting on the microstructures are transformed and thus ensure the requisite harmony of interactions between organic solids and liquids.

Furthermore, if the processes of development and growth imply specific mechanical operations that modify and amplify the original microstructures, contingency, by virtue of the organic mechanism at work, shows up in the effects resulting from these processes. Anomalies, monstrosities, disease, and even death can be explained by the modalities of movement affecting, respectively, the seeds, the preformed structures of the embryo, and the emerging organic constitutions.

It can be assumed that the seeds combine the purest extracts of the parent organic bodies; when they mix, they determine, in a differential manner, the insertion of specific molecules in the corresponding structures of the embryo. Thus, we can outline a representation of the transmission of hereditary traits that would spread to a plurality of phenomena observed through breeding between individuals and through the production of successive generations. The basic mechanism is explained as follows:

> The spirituous extract supplied by the father and mother is composed of particles as different as are the organs from which they originate; just as the fluids of the bodies of all animals are a mass of molecules of many species, which means that, by circulating within organs, they acquire strongly dissimilar properties. [...] It is assuredly the reason why the molecules contained in sperm, and which are prone to act [for example] in the organs of the face, determine more or less whether these parts will become similar to those of the father or the mother, as long as the infinitesimal parts of the embryo find themselves in a state that allows them to respond appropriately to this determination [...]. (Bourguet 1729, 153–54)

---

[22] See Malphighi (1673), Adelmann (1966), and Bernardi (1986, esp. 77–92).

Yet, such a determination can know multiple variations, according to the modalities of movement affecting the configuration of the organs that constitute the global mold, the transmission speed of the nervous influx represented by animals spirits, the composition and dynamic properties of the seeds, and the state of the embryo at the moment of development. The prevalence of certain parental resemblances, particularly illustrated in hybrid formations (Bourguet 1729, 161–64), the heredity of ancestral traits, and the morphogenetic impact of the imaginative representations of the mother on the foetus during pregnancy are notably linked to the diversified course of spirits according to sensual impressions, and linked to the resulting physiological effects in the phases of development and growth.

If Bourguet links animal functions to the expression of a psychical agent that, in the case of a foetus, appears to be "an active principle that acts without knowledge, responding by its spontaneous action to that of the agitated organs" (Bourguet 1729, 159–60), it's that he supports the Leibnizian thesis of a regulated expression between physiological determinations and the perceptive and appetitive states that seem to be their efficient causes. In that spirit, all the processes that can be linked to sensual perception—be it considered conscious or infra-conscious—and to the imaginative representations that seem to ensue are reduced to strict mechanical analogs: "only organic mechanism can produce such marvels, because it makes organized bodies communicate with the world at large, and by the means of these bodies, it makes the world of immaterial souls communicate with all the creatures of the universe" (Bourguet 1729, 160). The methodological inference that can be gleaned from this position consists in requiring a strictly mechanical explanation of the phenomena of sensual communication that are established between the sphere of ambient physical determinations, the morphological and physiological state of the mother, and the development of the foetus. The same requirement of reduction to organic mechanism also applies for functions operating within organized bodies that are progressively developed and moved. And so it is with nutrition, which rests on processes of assimilation and disassimilation realized within networks of fibrillar micro-canalizations. And so it is with the spontaneous movement of animals that we must account for with special mechanistic models of the type that Giovanni Alfonso Borelli (1608–1679) proposed in *De motu animalium* (1680–1681).

In that perspective, the reference to dominant and subordinate monads serves to symbolize the dynamic integration principle that ensures the structural and functional harmony of organized bodies. The order of active principles is reflected in the architectonic arrangement of preformed minimal organisms and in the internal combination of macroscopic organisms stemming from small embryonic machines by a whole series of developments and transformations. Aside from the representation of the deep forces underpinning vital organization as analytically impassable data, the recourse to monads must step aside in scientific explanation in favor of mechanistic models appropriate to figuring the transformation sequences of the preformed living. "Let us conclude, wrapping up all that has been said so far," declares Bouguet, "that organic mechanism is nothing but the combination of the movement of an infinity of ethereal, airborne, aqueous, oleaginous, saline, terrestrial, etc. molecules suited to particular systems, determined from the beginning by the Supreme Wisdom and each linked to a singular and dominant activity or monad to which those that enter into its system are subordinate" (1729, 164–65).

6.4 Conclusion

The great preformationist essays of the early eighteenth century, notably Antonio Vallisneri's *Istoria della generazione dell'uomo e degli animali* (1721), illustrate a confluence of methodological and metaphysical theses. In these, the experimental approach of the micromechanists is linked to principles of interpretation of the system of nature that essentially derive from neocartesian philosophies, notably that of Malebranche. A strongly different theoretical trend, inspired by Leibnizian concepts of force, monads, and organic bodies, tends to favor a slightly different research program, oriented much more toward the dynamics of the development and transformation of primordial organic structures. If the original production of these integrated microstructures still appears irreducible to a purely mechanistic explanation by the combination of inorganic particles, and even organic molecules, all can and must be interpreted mechanically in regards to activation, metamorphoses, and the sometime radical alterations that embryonic germs are subjected to in the phenomenal processes of generation. In that perspective, the Leibnizian model of the living being that is articulated, after the invention of dynamics per se (1690), through new or renewed concepts, such as those of monad, organism, and organic body, exerts a notable influence. While the inspiration taken from Leibniz begins to manifest itself in Vallisneri, it is fully integrated in Louis Bourguet's *Lettres philosophiques sur la formation des sels et des crystaux, et sur la génération et le mechanisme organique des plantes et des animaux* (1729). He conceived of pre-existing germs not so much as exact miniatures of the organisms that would subsequently result from them but more as matrices of complex and integrated organization on which acts "organic mechanism" so as to complete and modulate its composition and functioning. Furthermore, he appropriates the Leibnizian conception of the chain of beings and understanding of the material organization of the body itself, which is seen as a vital machine comprised of integrated organs, as the expression of the inherent limits of monads or "active principles" that exert their actions within it. As for the theory of generation, Bourguet considers that the various hypotheses must be put to the test of explaining the most complete set of generic phenomena corresponding to this function. With this goal in mind, he invalidates the hypothesis of "organic molds" and that of "plastic natures" and he corroborates that of "developments" from preformed structures; furthermore, he integrates into this explicative hypothesis the salvageable elements of the two other systems. Organic mechanism assumes, of course, already organized minimal bodies, but it determines, by acting on such a basis, the processes of development, nutrition, growth, and functional and spontaneous movement by which the organism transforms and realizes itself. Furthermore, the reference to dominant and subordinate monads takes into account dynamic integration, which ensures the structural and functional autonomy of organized bodies.

References

Adelmann, H. B. 1966. *Marcello Malpighi and the evolution of embryology*. Ithaca, NY: Cornell University Press.

Bernardi, Walter. 1986. *Le metafisiche dell'embrione, scienze della vita e filosofia da Malpighi a Spallanzani (1672–1793)*. Firenze: L.S. Olschki.

Bourguet, Louis. 1729. *Lettres philosophiques sur la formation des sels et des crystaux et sur la génération et le mechanisme organique des plantes et des animaux à l'occasion de la pierre bélemnite et de la pierre lenticulaire, avec un mémoire sur la théorie de la terre*. Amsterdam: Chez François L'Honoré.

Andry de Boisregard, Nicolas. 1700. *De la génération des vers dans le corps de l'homme*. Paris: L. d'Houry.

Duchesneau, François. 1993. *Leibniz et la méthode de la science*. Paris, Presses Universitaires de France.

Duchesneau, François. 1994. *La dynamique de Leibniz*, Paris, Vrin.

Duchesneau, François. 1995. Leibniz et la grande chaîne des êtres. In *Nature, histoire, société. Essais en hommage à Jacques Roger*, eds. C. Blanckaert, J.-L. Fischer, and R. Rey. Paris: Klincksieck.

Duchesneau, François. 1998. *Les modèles du vivant de Descartes à Leibniz*. Paris: Vrin.

French, Roger. 1994. *William Harvey's natural philosophy*. Cambridge: Cambridge University Press.

Generali, Dario. 1987. Antonio Vallisneri "corrispondente" leibniziano. In *Rapporti di scienziati europei con lo studio bolognese fra '600 e '700, Studi e memorie per la storia dell'Università di Bologna,* M. Cavazza (a cura di). Nuova serie vol. VI. Bologna: Istituto per la storia dell'Università.

Hartsoeker, Nicolas. 1730. *Cours de physique*. La Haye: J. Swart.

Leibniz, G. W. 1965. *Die philosophischen Schriften*, hrsg. von C.I. Gerhardt, Hildesheim, G. Olms. (GP)

Malpighi, Marcello. 1673. *Dissertatio epistolica de formatione pulli in ovo*. Londini: J. Martyn.

Roger, Jacques. 1971. *Les sciences de la vie dans la pensée française du XVIIIe siècle. La génération des animaux de Descartes à l'Encyclopédie*. 2e éd., Paris: A. Colin.

Vallisneri, Antonio. 1721. *Istoria della generazione dell'uomo e degli animali, se sia da' vermicelli spermatici o della uova*, Venezia: Gio. Gabbriel Hertz.

Vallisneri, Antonio. 1733. *Opere fisico-mediche*. Venizia: S. Coletti.

Part III
Rethinking Spinoza

Chapter 7
"Nemo non videt": Intuitive Knowledge and the Question of Spinoza's Elitism

Hasana Sharp

> *Summus Mentis conatus, summaque virtus est res intelligere tertio cognitionis genere.*
>
> -Ethica, Vp25

Scientia Intuitiva in Spinoza's *Ethics* is often claimed to be incomprehensible and to belong to an elitist doctrine of rarefied knowledge available only to a small class of philosophers. Jonathan Bennett has famously identified it as one of the three doctrines in the final part of the *Ethics* that can only be considered "an unmitigated ... disaster" (Bennett 1984, 357). While several commentators have objected to Bennett's complete dismissal of the three final doctrines (intuitive knowledge, the intellectual love of God, and the eternity of the mind),[1] it is generally agreed upon that intuitive knowledge is one of the most difficult concepts in all of Spinoza's philosophy. At least two interpreters have proposed that, by virtue of being intuitive knowledge it cannot be an object of reason and, therefore, remains unavailable to discursive description and analysis (Floistad 1969, 60; Klein 2002, 300). The difficulty of grasping precisely what Spinoza means by it is one reason philosophers reserve any experience of such knowledge to themselves, and only the "happy few" philosophers at that.[2] It is as though commentators presume that, if cannot understand what Spinoza means by intuition, it must be a mystical or otherwise inaccessible phenomenon.

Yet, although Spinoza's words about intuition, also called "the third kind of knowledge," remain among the most difficult to grasp, I argue that he succeeds

H. Sharp (✉)
McGill University, Montreal, QC, Canada
e-mail: hasana.sharp@mcgill.ca

[1] Yovel (1989, 232–233 n.2), for example, notes that Bennett's remarkably impassioned objections betray that he is in fact quite affected by what he calls mere "rubbish which causes others to write rubbish," and it cannot therefore be completely dismissed.

[2] Yovel (1989, 154). The notion that intuition is exclusively reserved for the sage who is not only intellectually privileged but also solitary, protected from the influence of the vulgar, is shared by a number of commentators, including Smith (1997), and Jon Wetlesen (1979).

C. Fraenkel et al. (eds.), *The Rationalists: Between Tradition and Innovation*, The New Synthese Historical Library 65, DOI 10.1007/978-90-481-9385-1_7,

in providing an account of its distinctive character. Moreover, the special place that intuition holds in Spinoza's philosophy is grounded not in its epistemological distinctiveness, but in its ethical promise. I will not go as far as one commentator to claim that the epistemological distinction is negligible (Malinowski-Charles 2003), but I do argue that its privileged place in Spinoza's system belongs primarily to its ethical importance, by which I mean that intuition's value is prized by virtue of the agency it confers upon those who enjoy it. Spinoza often notes that what distinguishes the "wise man" is that he is more powerful, able to *do* more. That is, the wise and free person with whose description the *Ethics* culminates is set apart not, first and foremost, by virtue of possessing a scientific account of the true cosmological order. The *Ethics* is perhaps called an ethics, rather than a treatise on metaphysics or knowledge, because it concludes with a demonstration of how one comes to be able to do what is best and to live a more joyful life. Intuitive science is essential to the characterization of the power to think and live well.

In reconstructing an account of intuition's distinctive character, I contend that the doctrine should not be understood as elitist for two reasons. I understand the charge of elitism to consist not merely in a descriptive claim that the enjoyment of intuitive science is, like all things excellent, "difficult and rare" (Vp42s).[3] Rather, an elitist doctrine is comprised of the generic claim that, in all or most circumstances, the sage *ought to* avoid the many and that wisdom is and ought to remain a private virtue. I interpret the claim that ethical perfection is difficult and rare to assert that "sagesse," to borrow a French word, or wisdom and the agency to which it belongs, is difficult and rare for anyone, not that any society will include only a few sages whose tremendous virtue enables them to rise above the crowd. I base this understanding on the following textual evidence. First, the intuitive grasp of the mathematical proportional used to describe the three kinds of knowledge in *Ethics* II is something that no one fails to see (*nemo non videt,* IIp40s2). While it has been argued that the mathematical examples perhaps should only be taken to represent a "partial analogy" with the genuine intuitive apprehension of concrete particulars (Curley 1973, 30), I claim that the model of mathematics is essential to Spinoza's rationalism, albeit not in the way it is often taken to be. One reason Spinoza sits both obviously and uneasily within the rationalist tradition is by virtue of the prominent but peculiar role played by mathematical truths throughout his philosophy.

The second reason the third kind of knowledge should not be taken to comprise an elitist doctrine of exclusive this-worldly salvation for the wise and happy few is that the "higher" expression of intuition described in *Ethics* V is not the privilege of any human types, or class of persons, but rather depends upon an arduous transformation of character and intellect contingent upon myriad factors beyond the control of any individual, or class of individuals. The difficulty of experiencing intuition in its highest form is attributable to the fragility of the knowledge of *any* finite being

[3]I will proceed to cite Spinoza parenthetically in the body of the paper with the standard notation, using Edwin Curley's edition and translation (1985). Citations refer to the part (= Roman numeral), proposition (= p), demonstration (= d), scholium (= s), corollary (= c), appendix (= app), preface (= pref), and definition (= def). The Latin is from Benedict de Spinoza's works (1925).

amidst infinitely many powers greater than itself (cf. IVp3). If the difficulty of regular enjoyment of intuitive knowledge is attributable to our finitude, the possibility of its cultivation and enjoyment is nevertheless ontologically presupposed by the eternity of the mind, which is a universal feature of mental life. Therefore, the character of our lives as finite modes, radically dependent both on God and one another for our existence and action, makes the full enjoyment of intuitive knowledge both universally possible and arduous.

My argument rests partly upon my claim that we can distinguish between two "moments" of *scientia intuitiva*.[4] The first moment is one that everyone experiences, while the second one, described in the second half of *Ethics* V is, indeed, difficult and rare. Commentators who stress its unattainability pay attention only to the second moment, but acknowledgement of the first forces one to conclude that it is at least an overstatement to claim that intuition is scarcely available to the many. Moreover, if one disregards the first moment on the basis of its mathematical character in favor of the intuition of singular existent beings emphasized in the second moment, one loses the importance of the *experience* of mathematical verities as an alternative and salutary "norm of truth" (Iapp; my translation) essential to Spinoza's rationalism. Thus, much of the paper will comprise an argument for understanding intuition along a continuum of basic mental experience, on one end, and the ethical cultivation of agency ("virtue") in concert with ambient modes, on the other. One can thereby understand intuition as both universal and rare, as long as one grants qualitatively distinct experiences of the same phenomenon. Understanding intuition with respect to either moment entails including it as part of "the true good, capable of communicating itself," in every sense, sought by the narrator of the *Treatise on the Emendation of the Intellect*.

I will conclude, finally, with a very brief discussion of the many times that Spinoza expresses doubt about the abilities of individuals to achieve mental freedom and peace of mind. Although Spinoza expresses these worries most often about the common man, the multitude, or the crowd (*vulgus*), he explicitly affirms that "all men share in one and the same nature"[5] and that the more people that are included in decision-making bodies, the more likely it is that the resulting decisions will be rational.[6] In both the *Ethics* and especially the *Political Treatise,* however, Spinoza blames the difficulty of living wisely upon the way lives, commonwealths, and institutions are "organized" or "constituted." Thus, I will claim that the difficulty inherent in attaining a life of wisdom and enjoying the full benefits of *scientia intuitiva* belongs to the relative powerlessness of any finite individual rather than expressing some ineradicable tendency prevalent among the ignorant masses alone.

---

[4]By two "moments" I mean to identify two analytically separable, yet intrinsically related aspects of intuition. Malinowski-Charles (2003, 2004) similarly identifies two "models" of intuition.

[5]Spinoza (2000, Ch.7, par. 14). Hereafter, TP.

[6]See TP, Ch. 9, par. 14, and Spinoza (2001, 178). Hereafter to be cited as TTP.

7.1 *Nemo non videt*: Scientia Intuitiva, Part I

Within the inventory of knowledge types in Part II of the *Ethics,* the third kind of knowledge appears almost as an afterthought. The description of the second kind of knowledge begins by noting that "*finally*, [knowledge] from the fact that we have common notions and adequate ideas of the properties of things . . . I shall call reason and the second kind of knowledge" (IIp40s2, my emphasis). Reason appears to be the ultimate way of knowing. In his edition of the *Ethics*, Curley inserts a roman numeral, which does not exist in the Latin, to mark a distinction when Spinoza proceeds to say that "Beyond [*præter*] these two kinds of knowledge, there is (. . .) a third, which we shall call intuitive knowledge [*scientiam intuitivam*]" (translation modified). Given that Spinoza inserts a "finally" prior to his definition of reason, the third kind of knowledge appears to be either tacked onto the other two or meant as a mere extension of rationality.[7] Spinoza proceeds to define intuition in the following way, which he will repeat verbatim in Part V: "this kind of knowing proceeds from an adequate idea of the formal essence of certain attributes of God to the adequate knowledge of the formal essence of things."

Spinoza continues by providing the well-known example of three ways of arriving at the fourth number given in the following mathematical problem: Find the fourth number with the same ratio to 3 as 1 is to 2. Spinoza proceeds to illustrate how one can arrive at the correct number using each of the three kinds of knowledge. In this example, the three kinds of knowledge do not pertain to different objects, but rather to different relationships toward the same object. Each way of knowing eventually arrives at the correct fourth number, is capable of being true, and yet designates a different means of reaching the same conclusion. Upon describing the third kind of knowledge, Spinoza remarks that in the case of this simple proportional, *no one* fails to *intuit* the right answer. Thus, this phenomenon, notorious for being so "difficult and rare" in the fifth part of the *Ethics,* is universally accessible vis-à-vis a simple mathematical proportional. "Given the numbers 1, 2, and 3, *no one* fails to see [*nemo non videt*] that the fourth proportional number is 6—and we see this much more clearly because we infer the fourth number from the ratio which, in one glance [*uno intuito*], we see the first number to have the second" (IIp40s2; my emphasis).

Only by regarding the mathematical example as irrelevant can commentators continue to consider the third kind of knowledge to be a highly elusive and even mystical doctrine, experienced as both private and incommunicable, irreducible to either signification or rational discourse, and available only to the few. In this

[7] Several commentators treat rationality as inclusive of the second and third kinds of knowledge, rendering intuitive science a kind of sub-species of reason. This is partly due to the fact that Spinoza indifferently applies the term "intellect" to both kinds of knowledge and because both kinds consist in "adequate ideas." While it is important to acknowledge the "ethical superiority" of the third kind of knowledge, I share the view that there is meaningful continuity between what is called reason and intuition. Moreover, it is appropriate to identify both as proper to Spinoza's peculiar rationalism. I will clarify my understanding of the ethical virtues of intuition below.

example, the intuitional inference is something *no one* fails to make, and the fourth proportional is seen clearly in one act of the mind. Nevertheless, commentators legitimately puzzle over the fact that, upon providing the same mathematical example (with some variation in the account of knowledge) in the *Treatise on the Emendation of the Intellect,* Spinoza notes that "the things I have been able to know by this kind of knowledge are very few."[8] It would seem that mathematics would offer a great many "things" to be known intuitively, and the impossibility of not knowing simple mathematical truths in this way seems to entail a highly accessible experience of intuitive knowledge. Let us first interrogate further how intuition is characterized in *Ethics* II before determining the status of the mathematical example.

Spinoza establishes some differences between the third kind of knowledge and the other two. He immediately indicates a couple of them: "Knowledge of the first kind is the only cause of falsity, whereas knowledge of the second and third kind is necessarily true" (IIp41). Both reason and intuition are considered "adequate ideas." An adequate idea "considered in itself, without relation to an object, has all the properties, *or* intrinsic denominations of a true idea" (IIIdef4). While the doctrine is complicated and a full account of it is beyond the scope of this paper, I will note that the truth of an adequate idea is not validated by apprehending its correspondence to an external object—that is, it is not true by virtue of grasping it insofar as it represents the world of the extension—but rather depends upon a relationship to its cause. An adequate idea includes the idea of its cause. By virtue of including their proximate cause adequate ideas can only beget adequate ideas, and thus the procession from an adequate idea—be it of common properties (reason) or the essences of certain attributes (intuition)—is necessarily true. Therefore, intuition, along with reason, is distinguished from imagination by virtue of its necessary truth, its ontological status as adequate idea.

Spinoza follows this proposition with what might seem to be an unnecessary addition: "Knowledge of the second and third kinds, and not of the first kind, teaches us to distinguish the true from the false" (IIp42). Whereas the first kind of knowledge can happen to be accurate, as in the example of a merchant who imitates what someone has shown him in order to perform a calculation, it does not have the *force* of a true idea, and it cannot thereby *teach* one to distinguish between what is true and what is not. The merchant is satisfied by the idea that the fourth number in the series is 6 by virtue of his faith in his teacher, an external authority. The first kind of knowledge is largely sufficient for daily functioning, but the idea that explains its truth is not internal to one's mind. Most importantly, imagination does not carry with it the *feeling* of certainty (even if Spinoza worries that it often includes the passion of conviction, as with superstition). "He who has a true idea at the same time knows he has a true idea, and cannot doubt the truth of the thing" (IIp43). As Spinoza emphasizes in the *Ethics* as well as the *Treatise on the Emendation of the Intellect,* truth must be its own standard. If something outside a true idea is

[8] Also in Spinoza (1985, §22).

required in order to measure its veracity, one will be met with an infinite regress. A true idea is a compelling force by which the mind experiences its own power, by which the mind experiences itself as the "adequate cause" of its own idea (IIIdef1), and is thereby unable to doubt it. When one intuits, in one glance, the fourth number in the proportional, the undergoing of the force of this act suffices to indicate to the knower that her idea is true. This very experience of certitude becomes the teacher, the standard, by which one distinguishes between the true and the false. At a minimal level, there is "no one" who has not intuited the most basic proportions, the most fundamental structural relationships, and, therefore, *every one* has some measure—an internal criterion or experience—by which to distinguish true and false ideas.

Mathematics appears elsewhere in the *Ethics* as a teacher. Moreover, unlike the teachers of merchants, mathematics offers the "force of demonstration" by which one can understand the causes of ideas and thus inspires "the eyes of the mind" to see their truth for themselves. In the appendix to *Ethics* I, for example, Spinoza outlines the universal tendency to imagine God or nature as a reflection of one's own idiosyncratic disposition (*ingenium*). Our native ignorance of causes and experience of our own appetites disposes us to apprehend nature or God as capricious, goal-oriented actors like ourselves. Humans tended to be guided primarily by their particular prejudices until they arrived at "another standard [*normam*] of truth." The truth that final causes in nature are nothing but human fictions would have remained "hidden from the human race to eternity, if mathematics, which is concerned not with ends, but only with essences and properties of figures, had not shown man another standard of truth" (Iapp).

Curley, for example, aims to diminish, or at least exercise utmost caution with respect to the mathematical examples, for fear that exclusive attention to them supports the erroneous image of Spinoza as an extreme rationalist who denies the importance or validity of human experience altogether. Curley is undoubtedly correct to combat the notion that Spinoza is a radical rationalist who aspires to a science on the model of pure mathematics, cleansed of all human passions, partiality, and lived experience.[9] To call Spinoza a rationalist, in this sense, according to Curley, "is not just mildly inaccurate, it is wildly inaccurate" (1973, 26). I agree. Yet, his wariness of the role of mathematics is likewise unwarranted, since Spinoza is clear that, far from providing *the* standard of truth, mathematics offers "another standard of truth." Mathematical knowledge is not meant to replace experiential knowledge, and neither is it to become the paradigm of any and all rationality. The dichotomy between mathematics and experience is a false one, for Spinoza. In fact, I would like to argue that mathematical knowing is valuable precisely *as an experience* of a different way of knowing. Mathematics offers the feeling of a verity grounded not in one's appetites with respect to empirical objects, but rather grounded in the pleasure

[9] While this may sound like a caricature, such an interpretation of Spinoza is not rare and is shared by a number of his detractors (and even some of his admirers). Two detractors in this line of thought include Alquié (1981) and Nussbaum (2001).

the mind takes in experiencing its own power. When one thinks in terms of the model of mathematics, one considers things with respect to their proper definitions, or essences, and not primarily with respect to how they affect the self. One apprehends that a triangle's angles, by definition, must add up to 180°. While one presumably assents to this truth independent of one's feelings about triangles, one is *affected* by the pleasure of contemplating something as it necessarily is. Since the mind is characterized by a desire to posit its activity, mathematics is an often simple way to enjoy the power to generate adequate ideas. Likewise, it has the added benefit of providing an alternative model with which to engage the often turbulent and messy human world. Thus, Spinoza encourages his reader, famously, to "consider human actions and appetites just as if it were a question of lines, planes, and bodies" (IIIpref). Thinking of things in terms of what follows necessarily from their essences, as one does with mathematics, teaches one to consider even the most painful human phenomena, not as essentially mathematical, but as analogous to mathematical truths. Mathematics, not as the truth of being but as a model of knowing, has an ethical function in that it enables us to cultivate an ability to affirm that "men, like other things, act from the necessity of nature." If we can do this, "the hate usually arising from [encountering the frequent wrongs of men] will occupy a very small part of the mind" (Vp10s). If, as I will discuss further below, the ethical superiority of intuition lies in its greater affective power, the mathematical example of this kind of knowing should be viewed in terms of the affects it generates and makes possible. If one illuminates this feature of the mathematical model, one can affirm that Spinoza's rationalism privileges mathematics without in any way diminishing experience, or even human emotions. Mathematics, in other words, is valuable not because it is a way of knowing independent of experience, but rather because it is a particular experience of knowing.

Before proceeding to treat the ethical superiority of intuition, I must discuss what it means to consider things in terms of their essences, since this is arguably the important epistemological distinction between reason and intuition. My argument is not that there is no epistemological difference between intuition and reason. Rather, I am claiming that the privileged place that intuition holds in Spinoza's thought is based on its ethical superiority. From the point of view of truth, reason and intuition are peers. From the perspective of power, as I will argue below, intuition is superior. Yet, the power of intuition arguably flows from its ability to grasp singular essences rather than common properties. Thus the epistemology is not as distinguishable from the ethics, as my formulation makes it seem. Nevertheless, I think it is worth making the heuristic distinction between the truth and practical values proper to ways of knowing. I will thus proceed to outline the peculiar epistemic character of intuition before arguing for its ethical priority.

Whereas reason is grounded in adequate ideas of "common notions"—what is shared among bodies—intuition sets off from an adequate idea of common *essences* to an adequate idea of singular *essences.* Reason generates ideas from basic rules and universal properties, but intuition operates from an apprehension of something universal or common (essences of certain attributes of God) to something singular (the essence of a thing). Everything that the human mind perceives is a modification of the attribute of thought or of extension. Nothing falls outside of a relationship to an attribute of God and, therefore, nothing exists that cannot, in principle, be intuited.

Attributes, or divine essences, are particular ways in which nature exists, particular forms of nature's power and self-expression. They are the determinate force that attributes to other beings their character as ideas or bodies. These essences, or powers, are intrinsic rather than extrinsic determinations of beings. Thought and extension are not "common notions" or properties existing equally in the part and the whole, like motion and rest, because they do not exist by virtue of the extrinsic determination and heteronomy that belongs to interaction among finite modes.[10] Attributes follow from the definition of substance, God, or nature. They are pure activity, pure attributive essence, necessarily flowing from infinite being, and giving their way of being to their modifications, or modes. There is nothing which is not a modification of an attribute of substance, and therefore attributes are common to all beings; that is, attributes are ontologically universal.

From an adequate idea of the essence of certain attributes of God—that is, from an adequate idea of this substantial activity constitutive of bodies or ideas as such—the human mind proceeds to an adequate idea of the singular essence of a thing. The essence of any given thing is its *conatus,* its effort to persevere in its being (IIIp7). Because human beings only conceive the attributes of thought or extension, they proceed from the essence that is thought, or the essence that is extension, to the essence of a particular idea or body. One proceeds from an adequate idea of a common power to a singular instantiation of that power, insofar as it strives to persevere in its being. In contrast to either the first or second kind of knowledge, we have left the realm of existence and are following a chain of essences. We are operating in the realm of a being's power insofar as it is intrinsically determined and remains *in Deo.*

Intuition begins from the basic proposition already affirmed by reason: "Whatever is, is in God, and nothing can be or be conceived without God" (Ip15). Essence is defined very similarly to this fundamental proposition of the *Ethics:* "that without which the thing can neither be nor be conceived, and which can neither be nor be conceived without the thing" (IIdef2). No being can be or be conceived without God, but the definition of essence adds the notion that "the thing" must also be given in order for the essence to be conceived. Thus, we do not conceive pure essences, or pure divine expressive power, without things that persevere in singular, distinctive ways. The third kind of knowledge includes an adequate apprehension of *natura naturans* and the conception of a singular, finite expression of the infinite power of thought or extension. Intuitive knowledge thus, concomitantly, apprehends the thing, its unique striving (*conatus*), and the primordial source of its striving.

---

[10]"Every singular thing, or any thing which is finite and has a determinate existence, can neither exist nor be determined to produce an effect unless it is determined to produce an effect by another cause, which is also finite and has a determinate existence; and again, this cause also can neither exist nor be determined to produce an effect unless it is determined to exist and produce an effect by another, which is also finite and has a determinate existence, and so on, to infinity" (Ip28).

This description, as abstract as it is, may seem applicable exclusively to living beings, beings we recognize as endeavoring to persevere in existence. Yet it includes mathematical knowledge as striving ideas. *Conatus,* essence or striving, does not apply only to human or living beings. Any and every mode in nature exerts a kind of effort to be what it is, to distinguish itself against the background of other modes, to be among the infinitely many diverse ways that nature exists. Ideas, no differently from bodies, strive to exist and thrive within an infinite force field of other ideas. Our minds are ideas, which aim to fortify themselves through producing increasingly many ideas. Mathematical ideas, like all other ideas, are not mute, dead, or passive reflections of some independently existent truth. They are endeavoring parts of our minds' efforts to exist and act effectively. The essences of triangles, albeit in a different way than biological organisms, should be understood to strive and act as modes within the attribute of thought.[11]

From the perspective of the second kind of knowledge, however, one does not begin with essences, or the activity of striving. Although the knowledge is likewise adequate, the order of reason's discovery reverses the order of being. From an experience of things in the world, we are able to deduce that they are all in God and all contain certain basic characteristics. As I noted, this procedure is not mistaken, and the second kind of knowledge remains necessarily true. We require the conception of things, or particular modifications of nature's power, in order to have the idea of nature at all. In the moment of intuition, however, the idea grasps the genetic metaphysical order by proceeding from the intrinsic and infinite power of nature to its singular instantiation, from universal essence to singular essence.[12] It grasps, then, not only *that* everything is in God, that every being involves and expresses the infinite power of nature, but *how* a singular thing is in God.[13] It grasps the finite singular as a modification of the infinite, and, at the same time, the particular limits of its power—the thing's essential striving to remain indefinitely what it is and to amplify its powers within the constraints of its peculiar nature. To apprehend something's essence, one must grasp, concomitantly, its source of power and its singular constitution, which entails some sense of its particular limits. One must apprehend it as a modification of the infinite self-affirmation of nature and as differentiated from every other modification. Thus, the third kind of knowledge proceeds from

---

[11]I develop this somewhat "vitalist" account of thought and ideas in Sharp (2007).

[12]Carr (1987, 246) argues that intuition's superiority to other kinds of knowledge belongs to it by virtue of its being correctly ordered.

[13]For a different approach, see Sandler (2005). Sandler shares Carr's view that reason and intuition are distinguished by the order of cognition rather than the object of cognition, but he rejects my claim that one actually has knowledge of a finite being's essence on the grounds that this involves improbable cognitive feats. Although I cannot justify this claim here, his argument that intuition must be restricted to knowledge *that* rather than knowledge *of* rests on his insistence that ideas are representational in nature for Spinoza. Sandler denies that one might represent to oneself, for example, the precise proportion of motion and rest that comprises a body's essence. A representation of m:n for mode X, however, would still remain at the level of imagination. Intuitive ideas, strictly speaking, are not representations of essences, but intellectual and affective apprehensions of them in terms of both their causal source and power.

the infinite "efflux" of substance (the attributes) to the finite intensity of singular beings.[14] It proceeds from what is concretely universal in all beings to what distinguishes them from anything and everything else. It is the single apprehension of both universality and singularity. It is universality and singularity at once, as the intrinsic power by which beings strive to persevere in existence, strive to express in their determinate way the infinite power of nature.

It is not difficult to imagine how one could do more with such knowledge. Individuals, of course, have their being in nature, and their strength comes from a common power to persevere in existence. They depend, however, not only upon God, but upon the infinitely many singular modes with which they necessarily interact. Individual strength and power emerges out of, rather than against, common power and striving, be it infinite or finite. There is no conflict between the singular and whole, in Spinoza's portrait of existence. The singular being is in the whole, and thereby strives to persevere *in* being. The third kind of knowledge apprehends precisely this co-existence of the common power (essence) of nature and the singular power to be what one is, irreducible to any other particular organization of that power. Intuition sees the singular power emerging out of the common, and what follows from it. It apprehends the essence of the singular as a productive, constitutive, causal force in and of nature, or God. As we will see in the following section, the full expression of intuition entails knowing increasingly many singular beings in this way, and thus enabling the knower all the more to interact with them in mutually beneficial ways.

To return briefly to Spinoza's example in *Ethics* Part II, the third kind of knowledge sees "much more clearly because we infer the fourth number from the ratio which, in *uno intuito,* we see the first number to have the second" (p40s2). We see clearly, because *from a relationship*, from a ratio, we see that the first *has* the second. The ratio is in a relationship of immanent belonging, and when that relational structure is "seen," we, easily and with the full feeling of certainty, grasp what follows from it. Relationships are causal. They bring things into being, like parents generating offspring, interlocutors producing ideas, or chemicals inciting reactions. Thus, the ratio of two terms—the modal essence within the attribute—entails a self-reproducing proportion, a kind of conceptual activity that continues to produce effects through the relational structure of ideas in which we are all implicated. On some level, however inchoate, "no one" fails to apprehend the ratio and what it involves.

7.2 Intuitive Superiority: Scientia Intuitiva, Part II

What distinguishes intuition, however, is not merely its epistemological power to apprehend being as it is, as something that flows from the infinite to the finite. Rather, as others have recognized, intuition is related to beatitude, the apex of human

[14]Spinoza asserts, in Letter 12, that modes cannot be understood if they are separated from substance and "the manner of their efflux from eternity."

freedom, and the power to organize one's affects in the most enabling way. That is, intuitive superiority is related to the height of ethics, where ethics is understood as the maximization of one's power (*potentia*) to think, feel, and act in and as a part of nature. Whereas the example from part II of the *Ethics* suggests that intuitive science is a relatively banal act that "no one fails" to accomplish, the second half of part V portrays it as a source of great power, dependent upon the ability of the sage to "remedy" her affects. While I mention the remedy for the affects, I do not analyze it since it merits its own discussion. I confine myself here to a discussion of the propositions naming the third kind of knowledge (Vp25 – p28, p31 – p33, p36s, and p38) in the attempt further to furnish a description of its distinctive characteristics.[15] In this section, I aim to describe the third kind of knowledge as it appears in part V, and in what its superiority consists.

Spinoza identifies several aspects of the superiority of the third kind of knowledge. Firstly, he notes that the "greatest striving of the mind, and its greatest virtue is understanding things by the third kind of knowledge" (Vp28). Virtue, for Spinoza, is coextensive with power:

> By virtue and power I mean the same thing, that is (by IIIp7), virtue insofar as it is related to man, is the very essence, or nature, of man, insofar as he has the power of bringing about certain things, which can be understood from the laws of his nature alone. (IVdef8)

"The greatest striving of the mind [*summus mentis conatus*]," the mind's essential activity, its constitutive desire and intrinsic drive is to cultivate its proper "virtue" or power. Due to the constraints imposed by the doctrine of the attributes, such that ideas can act only on other ideas, the power that belongs to the mind is nothing other than understanding. In contrast to a Cartesian morality where the mind aims to master the unruly passions of the body,[16] Spinoza holds that the mind's desire, properly understood and enjoyed, is strictly confined to the optimal augmentation of its particular agency within the realm of ideas. The summit of its own self-expression, the affirmation and cultivation of its own power as a modification of the attribute of thought, is *scientia intuitiva,* the third kind of knowledge.[17]

Spinoza proceeds to demonstrate proposition 25 by repeating the definition of intuition from part II, and asserting that "the more we understand things in this way, the more we understand God." Because we apprehend the essence of things from the essence of a divine attribute, and because everything is in God, we understand God more as we understand more singular things intuitively. This assertion requires that God be internally differentiated. Understanding the essence of God in the form of attributes, which are ontologically universal, does not provide the human mind with exhaustive knowledge of God or nature. Rather the mind understands more of God's power by understanding more singular differentiations, or modes. The more

---

[15] I do not address the vexing questions of the eternity of the mind and the nature of blessedness. On eternity, I can think of no better account than Jaquet (1997).

[16] See, for example, René Descartes (1985).

[17] The doctrine of the attributes, however, likewise entails that the power of the mind is the power of body and each is equally and concomitantly augmented by the development of the other.

I know about my peculiar striving to persevere in being and that of ambient modes, especially those with the greatest impact on my own power (e.g., loved ones, institutions that immediately affect me, etc.), the more I know God. God's essence is not uniform. Rather, divine essence is a productive power that yields infinitely many singular variations, infinitely many unique organizations of its power.[18]

The next proposition follows naturally: "The more the mind is capable [*aptior*] of understanding things by the third kind of knowledge, the more it desires to understand them by this kind of knowledge" (Vp26). Spinoza demonstrates this proposition with a characteristic remark: "This is evident. For insofar as we conceive the mind to be capable of understanding things by this kind of knowledge, we conceive it as determined to understand things by the same kind of knowledge." One ought to note that the capacity or aptitude of the mind emerges not by virtue of freedom from determination but the contrary. As one knows intuitively, the force of those ideas serves as a kind of momentum that cannot but yield more ideas of the same kind. The mind's power is activated by knowing in this way and continues to produce ideas that "follow from its nature." Since Spinoza often uses "nature" and "essence" interchangeably, to say that from this knowledge springs the "desire" to know still more from it is tantamount to saying that from one's (the mind's) essential nature follows its own nature. While this sounds tautological, the idea is that, with intuition, the mind better coordinates itself with external, ambient nature such that it can most effectively and ably amplify its striving, and thereby be what it essentially is. Moreover, this "desire [*cupit*]" that is the mind's essence may be called a determination (rather than, for example, a volition) for two reasons. First, the mind is determined by its very nature, its particular constitution as a complex mode of thought, to strive to understand as much as its nature allows. The mind aims to reproduce its joyful, or enabling experiences of knowing. Second, one may think of such knowledge as determined, or caused, because it depends upon the cultivation of a milieu of encounters, or relationships with ambient beings that enables it to maximize its power of understanding. This is evident if one recalls Spinoza's definition of essence: "the power [*potentia*] of each thing, *or* the striving [*conatus*] by which it (*either alone or with others*) does anything, or strives to do anything ... is nothing but the ... actual essence of the thing" (IIIp7d; emphasis added). Beings, even in their essence, act and strive "either alone or with others." Essences, then, often include the concurrent forces of other beings that enable one to act. A complex singular being cannot be said to have a discrete, isolable agency, or source of activity. The power of the mind depends not only upon recognizing, for example, the proper character of the universe, but also upon having the kind of corporeal and affective disposition that enables the mind to be ever more determined by its own pleasure of understanding than by anxiety, jealousy, hatred, greed, or other affects that tend to preoccupy the minds of social creatures like ourselves. Thus, Spinoza emphasizes that, rather than experiencing the illumination of the truth of existence

[18] Cf. Macherey (1979, Ch. 3).

all at once, each of us must strive, little by little, to know more and more singular things, especially those that contribute to and comprise our very power to exist.

Since the desire to know intuitively involves the actualization of the mind's essential power, it likewise discloses the mind's essence to itself. As others have emphasized, intuition involves, perhaps most centrally, a kind of self-knowledge. *Ethics* II emphasizes throughout, however, that the mind only knows itself by way of bodily encounters and the affects that they entail. If knowledge of more and more singular things includes knowledge of things other than oneself, it is nevertheless the case that the better one understands how one is situated the more knowledge of others is concomitantly knowledge of oneself. As Spinoza notes in part II, "the ideas we have of external bodies indicate the condition of our own body more than the nature of external bodies" (p16cor). The knowledge of more and more singular things, therefore, must depend upon knowledge of one's own body. Likely, the first part of *Ethics* V emphasizes corporeal and affective knowledge, or the "remedy for the affects," since it must serve as a precondition of the full enjoyment of intuitive knowledge. Intuition is thereby primarily a kind of self-knowledge, which both depends upon and generates an increased capacity to know external things and God, or nature.

The proposition that follows in Part V affirms the notion that the third kind of knowledge is firstly a form of self-knowledge: "The greatest peace of mind [*mentis acquiescientia*] there can be arises from this third kind of knowledge" (p27, translation modified). Spinoza sometimes attaches the term "*acquiescentia*" to the mind (*mens* or *animi*) and sometimes to the "self" (*in se ipso*). "*Acquiescentia*" is a neologism of Spinoza's that is not easily translated. Commentators often claim that Spinoza advocates a kind of Stoic rejection of human emotions,[19] but, given his criticism of Stoicism in the preface to part V, *acquiescentia* should be considered to be a revision of late Greek and early Roman notions of inner peace and self-possession. Curley does not render it consistently throughout his translation (translating it sometimes as "self-esteem" and others as "satisfaction of mind"), and thereby does not treat it as a term of art.[20] The term contains the Latin "*quies,*" and is thus related to "rest." I understand it to describe a condition of being at peace with one's self, affirming one's nature as it is, a kind of acquiescence to the constraints of one's particular essence, which resembles a Stoic notion. Yet, at the same time, it names an activity and an increased power, by virtue of acting out of, and in harmony with such constitutive constraints. It involves the liberation from bondage that belongs to knowing "both our nature's power and its lack of power" (IVp17s).

In part IV of the *Ethics,* Spinoza claims that "Peace in one's self [*acquiescentia in se ipso*] is really the highest thing for which we can hope" (p52s). He defines

---

[19]See, for example, Nussbaum (2001).

[20]I prefer the most common French translation *apaisement de l'âme,* yet the English rendering of *apaisement* as either "appeasement" or "pacification" has such negative connotations that I was forced to use a substantive "peace" when I would prefer a word suggesting transition and activity. I will maintain the word "peace," which resonates with "peace" as "strength of mind" and "unity of minds" in the *Political Treatise,* or leave it in the Latin.

"*acquiescentia in se ipso*" as "a joy born of the fact that a man considers himself and his power of acting" (p52d). From intuition arises this joy by which a person affirms her own power to act, to bring about effects in the world. Intuition begets this greatest peace of mind, this joyful experience of considering one's own effective and affective power, as part of the infinite power of God or nature. It thereby does not represent an overcoming of affective life, as for Stoicism, but an activation of the most enabling affects, especially those related to the self-love that flows from affirming oneself as a singular force of nature, or a unique expression of God's power (*gloria,* in its rational form, Vp36s). Whereas in part IV Spinoza asserts that "*acquiescentia in se ipso*" is born from reason, "*mentis acquiescentia*" arises from intuition. Donald Rutherford suggests that Spinoza "must be speaking loosely" when he claims that the former is "the highest thing we can hope for," since it is clear that salvation is that highest thing, and it belongs not merely to reason but to the mind insofar as it is eternal (Rutherford 1999, 459). He, therefore, makes the novel and interesting suggestion that these terms must not be equivalent. Given that it is also of the nature of reason to conceive things "*sub specie æternitatis,*" however, I do not see why the mind could not apprehend its own eternity by means of reason (IIp44). Thus, the nearly parallel assertions about reason (IVp52) and intuition (Vp27) giving rise to an enabling and liberating satisfaction of the self as it reflects upon its own power of action raises a question, I would argue, not about the distinct affects or forms of joy, but rather about the nature of the distinction between the second and third kinds of knowledge.[21]

The demonstration to proposition 27 of part V only confirms the parallel formulation from Part IV: "So he who knows things by this kind of knowledge passes to the greatest human perfection, and consequently (...), is affected with the greatest joy, accompanied (...) by the idea of himself and his virtue." Reason and intuition both produce an "*acquiescentia*" by which the mind rejoices in the apprehension of its own causal power. In the case of the third kind of knowledge, however, this includes the assertion that the knower "passes [*transit*] to the greatest human perfection." In what does this transition consist?

Spinoza's formulation echoes his definition of "joy" from part III: "By joy, therefore, I shall understand ... that passion by which the mind passes [*transit*] to a greater perfection" (p11s). Joy entails passing to greater perfection and intuition entails the superlative claim that, with the greatest joy, one passes to "the greatest

[21] I do not explore the problem of the validity of the distinction between reason and intuition in this paper. Malinowski-Charles makes a strong argument for treating them as "in reality the same knowledge, but simply under two modalities" (2003, 142). While I am not prepared to join this provocative argument, I certainly recognize the continuity between reason and intuition, especially by virtue of the fact that they both name "adequate ideas," and thus have the same ontological relationship to the mind that enacts them. Yet, there remains the question of whether knowledge of essences entails a meaningful epistemological distinction and, even more importantly, Spinoza recognizes a significant affective difference between the experience of intuition and that of reason. The affective difference is important for recognizing something like the ethical superiority of intuition, which may or may not be detachable from any epistemological privilege.

human perfection." This moment of highest human perfection is notably characterized by "an affection" of the greatest kind. Later in part V, Spinoza asserts that the superiority of intuition lies not in the fact that it contains or reveals more truth than reason, but in the fact that it *affects* the mind more powerfully. Spinoza claims that he aims here to show

> how much the knowledge of singular things I have called intuitive, *or* knowledge of the third kind (see IIP40S), can *accomplish*, and how much *more powerful* it is than knowledge of the second kind. For although I have shown generally in Part I that all things (and consequently the human mind also) depend on God both for their essence and their existence, nevertheless, that demonstration, though legitimate and put beyond all chance of doubt, still does not *affect our mind* as much as when this is inferred from the very essence of any singular thing which we say depends on God. (Vp36s; my emphasis)

The advantage is thus that the intuitive knowledge of singular things involves a more powerful affect and "accomplishes" more than reason. Spinoza's notion of intuition, in contrast to the post-Kantian one, is not devoid of any sensible content.[22] Although if one describes this solely from the point of view of the attribute of thought, one observes that intuition involves a great affection, an action of the mind upon itself such that it passes to the greatest power or perfection, the body concomitantly enjoys its greatest perfection. Although the second half of part V seems to focus on intellectual perfection and the eternity of the mind, it ends with discussion of the body (see, e.g., p39), as if to remind the readers of the "parallelism" of mind and body and that beatitude involves the whole person. Likewise, the note to proposition 39 reminds the reader that perfection is a developmental process of becoming increasingly capable of accomplishing more. "In this life, then, we strive especially that the infant's body may change (as much as its nature allows and assists) into another, capable of a great many things and related to a mind very much conscious of itself, of God, and of things." Intuition, then, does not indicate an exclusively intellectual power and does not mark an exit from the realm of sensuous affectivity. Indeed, the consummate force of the joyful affection distinguishes intuitive from rational knowledge. The increased agency such joyful wisdom involves accounts for the ethical superiority of intuitive knowing.

Proposition 27 circumscribes the particular character of intuitive affection, which proposition 36 reformulates. As I mention above, the affection by which one passes to the greatest perfection and enjoys the greatest joy is *acquiescentia* of the mind. *Acquiescentia* is a name for the joy experienced upon the contemplation of one's own power of action. One passes to the greatest affection by rejoicing in her own ability to produce effects and act in the world. Love, defined in Part III, "is nothing but joy with the accompanying idea of an external cause" (p13s). *Acquiescentia* is the self-love, the immanent joy by which one affects oneself with pleasure by affirming one's essence. In other words, *acquiescentia* consists in rejoicing in one's causal power in and of nature. It can be genuinely understood as self-love, but only

---

[22]For a brief account of intuition from the seventieth to the early twentieth century, see Rotenstreich (1972).

on the condition that one re-conceives the nature of this "self." It is one's singular power understood as an instantiation of the infinite power in and by which all beings exist and act. By acting, one is, at the same time, expressing divine, or natural power in a particular way, and is always acting within and by virtue of a causal community of infinitely many other finite beings. Moreover, because the infinite world of modal interaction participates in the actual determination of the character of divinity or nature, the mind can grasp itself as genuinely constitutive of, to borrow a phrase, "the real movement" of things.

The highest joy and passage to the greatest perfection emerges from the affirmation of oneself as a causal agent, a real constituent of nature, whose power is determined intrinsically by virtue of being a modification of divine attributes, and extrinsically by virtue of the co-affection of ambient modes. According to Spinoza, intuition always involves, simultaneously, an apprehension of causal relationality, divine or natural dependence, and a feeling of pleasure. "Whatever we understand by the third kind of knowledge, we take pleasure in, and our pleasure is accompanied by the idea of God as a cause" (Vp32). He continues in the demonstration, after linking intuition again to *acquiescentia,* "this joy is accompanied by the idea of oneself, and consequently (by P30) it is also accompanied by the idea of God, as its cause." We are pleased that our actions and our essence express the essential activity of God or nature. Our actions please us, then, not because we consider ourselves to be the unique authors of events in the world, imposing our imprimatur upon existence, but insofar as we express something eternal, definitive, and real. We are pleased in that we take God to cause our existence, and in that we concomitantly constitute nature, or being. Such causal power is equally intellectual and corporeal. With intuition, we are related to a greater number of existent ideas just as we become increasingly capable of affecting a greater number of existent bodies. We become, mentally and physically, increasingly "capable of a great many things" (Vp39s).

The first section of this paper demonstrates that intuition should be considered an act by which the mind grasps a singular essence as an instantiation of the common power of nature. Intuition reveals no more truth than reason, but "can accomplish more" (Vpref) and affects the mind more powerfully. I would like to suggest that it is more powerful in two senses. Firstly, knowledge of one's own essence as a modification of the essence of divine attributes delivers an indication of the particular limits and quality of one's power to think and act. Moreover, as one comes to know more and more singular essences, singular intensities of force, one can act more effectively than if one generates ideas from the consideration of universal properties, as is the case with reason. Intuition yields knowledge of how other singular essences affect one's own. It potentially discloses which beings are enabling and which are destructive. Since our power depends not only on our intrinsic determination by God, but also on our horizontal determination by ambient modes, knowledge of more and more singular essences begins to illuminate how collective, composite bodies can be constructed so as to increase the capacity to determine oneself and constitute more definitively the shape of nature.

Secondly, intuition is more powerful than reason because it corresponds to opening the body and mind to being affected more determinately by other essences (cf. IVp38). Such a phenomenon is not really distinct from the one described immediately above, but rather names the receptive rather than active aspect of activity immersed within a causal network. Intuition involves the affirmation that one's ability to act and think effectively in the world depends upon coordinating powers—be they intellectual or corporeal—that are necessarily involved with one another by virtue of being modifications of a single nature. The greatest joy and pleasure for humans emerge from our finite singularity as an instantiation of the infinite power of nature, in relationship to which there is no absolute interiority, no reserve that remains untouched. Knowledge of oneself as fully permeated by the divine essence brings the greatest joy and the most satisfying love, which are affects that name an increase in agency. Spinoza's intuitive knower, however, is not dissolved or disintegrated by this constant infusion, as the mystical interpretation would have it, but rather singularized, rendered more autonomous (albeit in a pre-Kantian sense) by such an affirmation. As one comes to affirm oneself as ineluctably, essentially a determinate force of nature, one can direct oneself more effectively. One is not lost, but empowered to the maximum extent by appropriating one's power of striving through understanding one's unique character. Such appropriation requires the simultaneous sense of oneself as determined intrinsically to express the power of God with one's activity, and, at the same time, dependent upon the concurrent activity of other modes—ideas as much as bodies—in order to carry out particular actions.

Intuition, then, is the joyful knowledge of singular powers that flow from the common power of nature. It is the pleasant affirmation of difference within the common. It is the thrill of finitude infused with the infinite. It is the maximization of the power to think and to act by virtue of knowing the self as it really is, and not as we might like it to be. On my interpretation, one of the greatest marks of distinction of Spinoza's rationalism is its emphasis upon power and capacity over the justification of belief. The conclusion of the *Ethics* promises to demonstrate not merely the virtues of understanding the cosmological order and one's place in it, but of disclosing "how much more the wise man can *do*" (Vpref). An ethics conceives of knowledge as an activity rather than a reflection. Thus, knowing more is an index of the mental and corporeal action, a concomitant of "doing" more. The mind strives, then, not simply to represent, to serve as the speculum, or mirror, of divine creation. Rather, the mind, from the point of view of Spinozan ethics, yearns to compose increasingly more of being itself with its active production of ideas. Thus, Spinoza's major work is called an *Ethics* rather than a physics, metaphysics, or a treatise of man and nature *simpliciter*. Moreover, the highest kind of knowledge is not distinguished by virtue of its superior truth value, but rather by the greater mental and corporeal capacity it delivers. Spinoza's rationalism might be peculiar in that it is the affect and not only the content of knowledge delivered by reason and intuition that is "the greatest thing for which we can hope." Likely due to our contemporary division of philosophical labor, students of Spinoza's *Ethics* treat his rationalism in isolation from the ethics or politics. This has resulted in a portrait of Spinoza as an "extreme rationalist" who restricts this "hope" to the solitary and disciplined sage who has no passion other than the drive for truth. Spinoza scholars often presume that the "we" who can hope for this greatest experience of freedom is precisely "us" philosophers, and is strictly denied to the *vulgus,* or even the *multitude,* the collective as such. In the final section, I will offer some brief suggestions that aim further to dismantle the view of Spinoza's rationalism as one that belongs to the dispassionate scientist who extirpates all emotional, experiential, and sensuous elements from his quest for eternal truths.

7.3 Wisdom for the Many?

It is not mere self-admiration that causes philosophers to assert that the height of knowledge and human freedom is to be enjoyed only by ourselves. Spinoza often makes assertions that, despite the fact that rationality issues from human community and the shared striving to live and know well, "it rarely happens that most men live according to the guidance of reason. Instead their lives are so constituted that they are usually envious and burdensome to one another" (IVp35s). Likewise, he suggests that a rational, "free man who lives among the ignorant strives, as far as he can, to avoid their favors" (IVp70). Thus, although Spinoza emphasizes the importance of human community, the rational desire to join oneself to others (IVp18s), and the profound value of mutual friendship (IVp71),[23] he often pairs such remarks with a voice of caution about "most men" and "the ignorant." Famously, in the preface to the *Theological-Political Treatise* Spinoza invites philosophers to examine his argument, but discourages the superstitious *vulgus,* the emotional crowd that prizes religious doctrine over reason, from reading the text at all.[24] Indeed, it sometimes appears that, while the *Ethics* promises that, under the guidance of reason, "man is a God to man," emphasizing the potential for harmony and sociability, the political works seem to show how "men are. . .by nature enemies" (TP, Ch. 2, par. 14).

While the issue of Spinoza's attitudes toward his own potential readers and the many in general is too complex to treat adequately in these brief concluding remarks, I would like to indicate a few reasons to question the assumption that intuition is in principle reserved for philosophers, the solitary sage, or the happy few. First, there are no metaphysical reasons that intuition ought to be confined to an intellectual elite. Intuition is conceptually and intimately linked to beatitude, the eternity of the mind, and freedom in *Ethics* V. While a full account of the doctrine of mental eternity remains beyond the scope of this analysis, it is beyond doubt that all minds, for Spinoza, are eternal in substance (Vp22). Any mind is an eternal idea whose eternity is metaphysically co-extensive with its existence. That is, one does not *achieve* eternity in the way one might earn immortality by way of one's actions. The eternity one enjoys in beatitude for Spinoza has nothing to do with a theological notion of the immortality of the soul (more akin to sempiternity). In other

[23]See also Spinoza's Letter to Blyenbergh, 5 January 1665.

[24]Spinoza, TTP. Leo Strauss discusses the question of the audience of the *Theological-Political Treatise* in detail in his (1988, Ch. 5).

words, each and every being is always already eternal by virtue of comprising a part of the infinite power of God or nature, but one can be more or less conscious of such eternity. As Jaquet demonstrates, while *discovery* of the eternity of the mind follows from intuitive knowledge, it does not depend upon it ontologically. Rather, the dependence is the other way around. Intuition "is not the cause but the active expression of salvation," as Spinoza conceives it.[25] Thus, there is no metaphysical barrier to anyone cultivating the kind of relationship to oneself and the world that enables the highest expression of intuition.

Rather, the barriers to becoming the supremely capable wise person with a "body capable of a great many things" are primarily social, political, and environmental. Moreover, such obstacles to freedom and agency impinge upon everyone, whether one is learned, mature, and strong, or ignorant, young, and undisciplined. Certainly, a number of people are inhibited at various points in their lives from enjoying the height of their mental powers by virtue of youth, illness, or infirmity of various kinds, over which no one has any power.[26] In general, however, Spinoza affirms that "all men share in one and the same nature; it is power and culture that mislead us." Indeed, "everywhere truth becomes a casualty through hostility or servility, especially when despotic power is in the hands of one or a few."[27] Spinoza proceeds to advocate large deliberative bodies and the transparency of state policies. He suggests that people are kept ignorant not only by being terrorized, but also by being excluded from the activity of decision-making and learning the rationales behind them.[28] Thus, the *Political Treatise* aims to furnish the objective structural principles that would allow as many people as possible are able to act *as if* they were guided by reason, and thereby cultivate a disposition that can think and act well. "If it is to endure, a government must be so organized that its ministers cannot be induced to betray their [constituents'] trust to act basely, whether they are guided by reason or passion."[29] Providing the structural conditions that will habituate the ministers as well as the citizens to acting in the interest of the whole serves as an education for all actors as well as a buttress against unanticipated misfortune. Thus, it is not the case that the political writings represent a more critical appraisal of humans as natural enemies. Rather, they comprise an analysis of those circumstances and institutional conditions that engender either enmity or solidarity (acting as "one mind").

The ability of each of us to think and act well is fragile and vulnerable to the turbulent passions which can inevitably befall any of us by virtue of misfortunes such

[25] See Jaquet (1997, 105). I recommend this text as the most precise account of eternity in Spinoza.

[26] Spinoza repeatedly mentions the inability to control whether one has a sound body (e.g., TTP, Ch. 16, and TP, Ch. 2, par. 18). In the *Ethics*, he also mentions the Spanish Poet who suffers such a grave illness that he loses his memory and therefore ought to be considered a different person (IVp39s).

[27] TP, Ch. 7, par. 27.

[28] See also, TP, Ch. 9, par. 14.

[29] TP, Ch. 1, par. 7.

as illness, natural catastrophe, war, and broken hearts, to name a few. Such vulnerability belongs to our finitude and is ineradicable. Moreover, we are vulnerable not only to what befalls us as individuals, but to what arouses and plagues those who surround us. Our susceptibility to being affected by others at any and every moment of our existence is the source of both our power and its ineradicable fragility. As Spinoza notes in the *Ethics,*

> It is impossible for man not to be part of nature and not to follow the common order of nature. But if he lives among those who agree with his nature, his power of acting will thereby be aided and encouraged. On the other hand, if he is among men who do not at all agree with his nature, he will hardly be able to accommodate himself to them without greatly changing himself. (IVappVII)[30]

One "agrees in nature" with those with whom one "agrees in power," or striving (IVp32d). The extent to which one lives among those with whom one can join powers and combine agencies toward living and thinking well, one can become more and more powerful and enjoy intuitive knowledge. Even if one has cultivated great mental and corporeal power, if the political and social environment becomes very hostile to one's striving, one cannot avoid being changed.

Those who disagree with one's nature, however, cannot be described as a natural kind (e.g., many fellow humans disagree with any individual's nature), and should not be viewed as a characteristic of collectivities as such. On the contrary, Spinoza regularly expresses hope that the commonwealth will combine into a massive "union or harmony of minds,"[31] or that "all should so agree in all things that the minds and bodies of all would compose, as it were, one mind and one body" (IVp18s). Rather than betraying anxiety toward collectives as such, both the *Ethics* and the political treatises detail the principles by which vital and enabling collectivities might be formed. While it is often true that a society includes many ignorant, violent, and hostile people, "*their lives are so constituted* that they are usually . . . burdensome to one another" (IVp35s). Among those whose life activities promote the most violent passions and within a government that is not "so organized that its ministers cannot be induced to betray their trust," the wise will "avoid the favors of the ignorant." That is, the wise will strive not to become dependent for their mental or material well-being on those who cannot act according to reason. Such dangerous individuals, for Spinoza, include not only the uneducated and superstitious masses, but also the unconstrained and capricious state ministers, as well as merchants preoccupied with financial gain. When circumstances are hostile to mental and corporeal fortitude, which is very often the case for Spinoza at the time he wrote his philosophy, he recommends striving not to rely upon the forces of those who disable one's efforts to think and act. In such cases, he might, with Machiavelli, counsel those who can to rely on their own forces. Yet, humans "can hardly live a solitary life" (IVp35s) and "it is vain for one man alone to guard himself against all others."[32]

---

[30]For a detailed discussion of this passage, see Beyssade (1994).

[31]TP, Ch. 6, par. 4.

[32]TP, Ch. 2. par. 15.

While this section has been all too brief, I hope to have established, first, that there is no metaphysical reason that intuition cannot be enjoyed by each and every individual. Second, I hope to have suggested that the difficulty that attends any efforts to enjoy reason as well as the highest expression of intuition belongs to the universal condition of our finitude. Because each of us remains ineradicably finite, a part of nature, each of us depends upon ambient modes for a vital and powerful mind and body. The modes that impact our well-being are innumerable, but include those circulating in the environment, such as air and water quality, major events such as wars, famines, plagues, or bumper crops, as well as social relations and customs. It is thus, indeed, "difficult and rare" to enjoy the highest expression of intuition and the peace of mind it brings. Yet Spinoza's doctrine of intuition should be understood to be neither an elitist program nor a solitary pleasure. While Spinoza often qualifies his hope for the blossoming of mass wisdom and collective harmony, such a community of agencies is precisely the aim of his philosophy. While it may not be possible for everyone to enjoy the heights of wisdom, it is also the case that no one will enjoy it all of the time. Optimal conditions and maximal cooperation will enable many people to experience it. Even if intuitive science will remain difficult and rare within the course of individual lives, it need not be a scarce good within a population. Intuitive knowledge, as an expression of the apex of ethical power, includes love toward God, and "the more men we imagine to be joined to God by the same bond of love, the more it is encouraged" (Vp20).

References

Alquié, Ferdinand. 1981. *Le rationalisme de Spinoza.* Paris: Presses universitaires de France.

Bennett, Jonathan. 1984. *A study of Spinoza's* ethics. Indianapolis, IN: Hackett.

Beyssade, Jean-Marie. 1994. VIX (Ethique IV Appendice chapitre 7) ou peut-on se sauver tout seul? *Revue de Métaphysique et de Morale* 4: 493–504.

Carr, Spencer. 1987. Spinoza's distinction between rational and intuitive knowledge. *The philosophical review* 87 (2): 241–52.

Curley, Edwin. 1973. Experience in Spinoza's theory of knowledge. In *Spinoza: A collection of critical essays* ed. M. Grene, Ed. Notre Dame: University of Notre Dame Press.

Curley, Edwin. 1985. *The collected works of Spinoza.* Vol. 1. Princeton, NJ: Princeton University Press.

Descartes, René. 1985. *Passions of the soul.* In *The philosophical writings of Descartes* trans. R. Cottingham, R. Stoothoff, and D. Murdoch. Vol. 1. Cambridge: Cambridge University Press.

Floistad, Guttorm. 1969. Spinoza's theory of knowledge. *Inquiry* 12: 41–65.

Jaquet, Chantal. 1997. *Sub Specie Æternitatis: Etude des concepts de temps, durée et éternité chez Spinoza.* Paris: Editions Kimé.

Klein, Julie. 2002. "By eternity I understand": Eternity according to Spinoza. *Iyyun* 51: 295–324.

Macherey, Pierre. 1979. *Hegel ou Spinoza.* Paris: Maspero.

Mailnowski-Charles, Syliane. 2003. The circle of adequate knowledge: Notes on reason and intuition in Spinoza. In *Oxford studies in early modern philosophy*, Vol 1, eds. D. Garber and S. Nadler. Oxford: Oxford University Press.

Mailnowski-Charles, Syliane. 2004. Habitude, connaissance et vertu chez Spinoza. *Dialogue* 43(1): 99–124.

Nussbaum, Martha. 2001. *Upheavals of thought.* Cambridge: Cambridge University Press.

Rotenstreich, Nathan. 1972. Bergson and the transformations of the notion of intuition. *Journal of the History of Philosophy* 10: 335–46.
Rutherford, Donald. Salvation as a state of mind: The place of *Acquiescentia* in Spinoza's *Ethics*. *British Journal for the History of Philosophy* 7(3): 447–73.
Sandler, Ronald. 2005. *Intuitus* and *ratio* in Spinoza's ethical thought. *British Journal for the History of Philosophy* 13(1): 73–90.
Smith, Steven B. 1997. *Spinoza, liberalism, and the question of Jewish identity.* New Haven, CT: Yale.
Sharp, Hasana. The force of ideas in Spinoza. *Political Theory* 35(6): 732–55.
Spinoza, Benedict. 1925. Opera. 4 vols. Ed. C. Gebhardt. Heidelberg: C. Winter.
Spinoza, Benedict. 1985. *The collected works of Spinoza* Vol. 1. Princeton, NJ: Princeton University Press.
Spinoza, Benedict. 2000. *Political treatise* trans. S. Shirley. Indianapolis, IN: Hackett Publishing.
Spinoza, Benedict. 2001. *Theological-political treatise* trans. S. Shirley. Indianapolis, IN: Hackett Publishing.
Strauss, Leo. 1988. *Persecution and the art of writing.* Chicago: University of Chicago Press.
Wetlesen, Jon. 1979. *The sage and the way.* Assen: Gorcum.
Yovel, Yirmiyahu. 1989. *Spinoza and other Heretics.* Vol. 1. Princeton, NJ: Princeton University Press.

Chapter 8
Rationalism Versus Subjective Experience: The Problem of the Two Minds in Spinoza

Syliane Malinowski-Charles

There is a long tradition of commentators for whom Spinoza's rationalism, taken to the letter, entails that subjective experience amounts to simply *nothing*. The idea here is that experience is indeed given in a subject, but that the illusory nature of imagination, of the passions, and, in a nutshell, of any finite perspective, means that the self that is constituted by this truncated knowledge is itself truncated, and even that it is devoid of any real existence. In other terms, it would exist only insofar as it is also an adequate idea in God. This is where the problem of the two "minds," or of the two "I"s, arises. For "I," as the subject of my largely passive experience of the world, am also, in God, the objective reality corresponding perfectly to my body. According to Spinoza, the idea that God's infinite intellect contains of everything (i.e., of all his modes[1]) is namely nothing but the "objective reality" corresponding to the "formal reality" of the thing given in the different attributes.[2] For instance, in the case of humans, who are constituted by a body and a mind, the mind is nothing but the objective reality (or idea) of a body. And as Spinoza states at E IIP32, "All ideas, insofar as they are related to God, are true" (G II, 116).

But since God's intellect can only comprehend things adequately, no passivity or inadequacy can be found in the mind or in the "I" which is really in God. Should we hence believe that everything in us that corresponds to inadequate knowledge—what Spinoza refers to, among other things, as "random experience"[3]— is nothing

S. Malinowski-Charles (✉)
Bishop's University, Sherbrooke, QC, Canada
e-mail: scharles@ubishops.ca

[1]Cf. E IIP3; G II, 87, quoted in section I below. Citations to the *Ethics* (in the English translation of Curley) are given as "E" followed by the part in Roman numerals, then "P" for proposition followed by its Arabic numeral, "D" for demonstration, "S" for scholium, "C" for corollary, etc. These indications are followed by the volume and page number in "G," i.e. the reference edition by Gebhardt (1925). "KV" refers to the *Korte Verhandeling (Short Treatise)*, TIE refers to the *Treatise on the Emendation of the Intellect* (quoted according to the sections by Bruder), and "Ep." refers to the *Epistolae* (letters).

[2]See E IIP11-13 (G II, 94–6) and KV II, App. II, 14–15 (G I, 120). On this, see my section I.

[3]See notably E IIP40S2 (G II, 122) for the different grades of cognition. The first kind of knowledge, which Spinoza globally calls "imagination," includes both random experience (*experientia*

C. Fraenkel et al. (eds.), *The Rationalists: Between Tradition and Innovation*, The New Synthese Historical Library 65, DOI 10.1007/978-90-481-9385-1_8,

but a fleeting, *unreal* fantasy? What ontological status does subjective experience have in a rationalist system where everything that is outside reason merely seems to become nothingness? Experience cries out against this reduction, for it is precisely in states of passion such as love, anger, or fear that we tend to feel that we are more alive—somehow, that we are more *real*. Did Spinoza really push the rationality of his system to the point of denying that "I" am something at those moments, or that they are unreal?

This paper will look at the problem of the two minds—"mine," and the mind in God—in four different respects. In the two first sections, we shall examine how the problem arises at the cognitive level, and how intellectualist readings of Spinoza could find some legitimacy in the texts themselves. Against these interpretations, in the third part I will strive to understand whether imaginative and perceptual cognitions can be said to be in God in one way or another, which involves accounting for the "truncation" of errors. Finally, I will aim to show that it does make sense to hold that these two ideas or minds are both real in Spinoza, and I will explain in which sense it may be said that the second (inadequate) mind is a *part* of the first one. By using some features of the adequate idea corresponding to us in God, the mind implicated in our subjective experience is intrinsically related to it. The difference is thus not one of existence, absolutely speaking, but of durational vs. eternal existence (which, in turn, sheds some light on Spinoza's doctrine of the eternity of the adequate part of the mind).

8.1 The Absolute vs. the Subjective Mind

Spinoza's account of the subjective self or mind is intrinsically linked with his theory of inadequate ideas. Despite their capacity to form adequate ideas that gives them a sort of superiority over other beings in nature,[4] human beings themselves have minds that are for the most part constituted by inadequate ideas. Why "inadequate"? Because they are incomplete; because, to use Spinoza's terminology in E IIP29S, "the mind has only a confused and mutilated knowledge, of itself, of its own body, and of external bodies" (G II, 114).

My inadequate ideas constitute a mind that is necessarily different from the adequate idea in God also corresponding to "me," but from an adequate perspective. According to E IIP7 (G II, 89), "the order and connection of ideas is the same as

---

vaga), which one acquires by direct perception and memory (i.e., directly), and knowledge from signs (*ex signis*), which one acquires by hear-say and education (i.e., through someone else). The two others, which are adequate, are reason and intuitive knowledge.

[4]Non-human beings all have a mind corresponding to them, and we may hold that some of them have consciousness, but they are still inferior for Spinoza (see E IIP13S). For a discussion of consciousness in the non-human realm, see chap. VI of my book (2004, 129–44), and my yet unpublished paper "Spinoza on Animal Consciousness." In them, I have defended the view that any being capable of affects is also capable of consciousness, which applies to the whole organic level in differing degrees (very low in plants, very high in some animals).

the order and connection of things." This entails that for each event or thing in any attribute of God, an idea of it is also given in nature, i.e., in God. For as E IIP3 tells us, "in God there is necessarily an idea, both of his essence, and of everything which necessarily follows from his essence" (G II, 87). This idea is itself what Spinoza calls the "mind" (*mens*) of the thing: "The first thing which constitutes the actual being of a human mind is nothing but the idea of a singular thing which actually exists" (E IIP11; G II, 94). The demonstration and corollary show that this is not only so in the human case. Indeed, all justifications are given from general propositions and axioms that would apply to any being. We can thus take in a general sense what E IIP11C says specifically about the human mind:

> From this it follows that the human mind is a part of the infinite intellect of God. Therefore, when we say that the human mind perceives this or that, we are saying nothing but that God, not insofar as he is infinite, but insofar as he is explained through the nature of the human mind, *or* insofar as he constitutes the essence of the human mind, has this or that idea. (G II, 94–95)

The dice are cast, and Pandora's Box is open: Spinoza says that the finite minds are part of God's eternal, infinite intellect (which is the immediate infinite mode in the attribute of thought). The rest of the corollary has interesting specifications that we shall analyze in our third section, but for now, we may "come to a halt" as Spinoza advised his readers (immediately after this corollary, in E IIP11S; G II, 95), and try to understand fully the problem of the two minds posed here. If the temporal, and in great part, inadequate idea constituting my mind is also a part of the infinite, eternal, and adequate idea which is in God's intellect, then what am "I"? Am I not reduced to being an abstract entity, and if I am actually an adequate idea in God, how is it that I don't have knowledge of this "I" which must also be "me"?

The problem is even sharper, for a consequence of this view is that an idea or "mind" is necessarily given in God as a correlate to any being; i.e., everything is necessarily "animated" (in the traditional, Aristotelian sense of having an *anima* or mind). This is what E IIP13S makes explicit: "The things we have shown so far are completely general and do not pertain more to man than to other individuals, all of which, though in different degrees, are nevertheless animate" (G II, 96–7). There has been much discussion as to how to interpret the parallelism of the attributes, and the universal animism it implies. It would seem that this thesis is very hard to interpret in terms other than logical ones. Curley famously held that beings in the attributes other than thought (i.e., among others, in extension) are "facts" instantiating the "laws of nature," for which the particular ideas in the attribute of thought are particular expressions in the form of true propositions.[5] The laws of nature in extension are those of motion and rest (i.e., what Spinoza refers to as the infinite immediate mode), and those in the attribute of thought are mathematical laws and logical rules. This amounts to saying that the mind corresponding to my body is, in God, a true proposition about the state of this body. This logicizing of the idea that God has of us, i.e., of our mind as an adequate idea, may seem hard to accept,

[5] See Curley (1969, 118–143).

and has in fact been heavily criticized[6]; however, something along the lines of this interpretation seems difficult to avoid, and is indeed well justified by the texts.

A passage from the *Short Treatise* proves that we are, here, at the roots of Spinoza's early doctrine and initial intuition about reality (which also has led Curley to talk about a certain form of "materialism" in Spinoza,[7] in the sense of a certain primacy of the state of the body of which the mind is only the mental mirroring). In his second appendix, called "Of the Human Soul," Spinoza offers an account of the mind-body relationship which is illuminating in many ways for understanding his final doctrine in the *Ethics*, despite the fact that the simultaneity of mind and body, and the universal aspect of the parallelism, were far from being as clearly elaborated there as in the *Ethics*.[8]

> Here, then, we shall suppose as a thing proven, that there is no other mode in extension than motion and rest, and that each particular corporeal thing is nothing but a certain proportion of motion and rest . . . So this existing proportions' objective essence in the thinking attribute is the soul of the body. Hence when one of these modes (motion or rest) changes, either by increasing or by decreasing, the idea also changes correspondingly. For example, if the rest happens to increase, and the motion to decrease, the pain or sadness we call *cold* is thereby produced. On the other hand, if this increase occurs in the motion, then the pain we call heat is thereby produced. (KV II, App. II, sections 14–15; G I, 120)

I do not find that the *Ethics* uses the (Scholastic and then Cartesian) vocabulary of the formal vs. the objective essence or reality of a being in any such clear-cut way, although it does express the same view.[9] The *Ethics* says, notably in E IIP21D and S (concerning the idea of idea defining our self-consciousness), that "the mind is united to the body from the fact that the body is the object of the mind" (E IIP21D; G II, 109), and that "the idea of the mind, that is, the idea of the idea, is nothing but the form of the idea insofar as it is considered as a mode of thinking without relation to the object" (E IIP21S; G II, 109). I prefer however this passage from the *Short Treatise* because it is clearer than any other: our body is, for Spinoza, our formal reality, of which our mind is the objective reality. Since the body is only a certain proportion of motion and rest, it seems natural to interpret this proportion's "idea" in mathematical terms, a little bit like a given chemical formula corresponds to any given physical type of body (water as H_2O, for instance). Since we can likewise interpret the composition of a given existing body in terms of the quantity it has of

[6]See for instance the very interesting criticism of Balz's (1918) and Curley's (1969) attempts at "logicizing" the attribute of thought (Bennett 1984, 129) in Bennett (1984), chapter II, section 14: "Psychology and Logic," 50–54, and chapter VI, sections 31 and 32, 127–34.

[7]See Curley (1988, 74–7), chap. II section 10: "Spinoza's Materialism." See also my discussion of this question in Malinowski-Charles (2009), forthcoming.

[8]In fact, Spinoza moved from a Cartesian model of mind-body interaction to his own final doctrine of their identity in different attributes. On this, see the last section of my forthcoming paper on pleasure and pain in Spinoza (2009), as well as Jaquet (2005), and a yet unpublished paper by Tammy Nyden-Bullock (2007), 90–115 entitled "Parallelism *à la mode*."

[9]It should be noted that this vocabulary is also explicitly used in the TIE, e.g. at section 33 (G I, 14).

each kind of molecule and atom, it is natural to understand the mind as the "idea" (i.e., logical or mathematical expression) of this ratio.

This is obviously a very abstract view of the mind, and consequently, a very impersonal one. How such subjective states as pleasure and pain can be reduced by Spinoza in this Appendix to the idea of a particular ratio of motion and rest among the composing parts of the body is not only fascinating, it is also troubling. It is no surprise, then, that the reconciliation of this abstract, merely logical (and always true) "idea" of us in God, with the one which is the subject of our daily, psychological, experience, has been seen as an impossible task by many. This problem has actually led a whole tradition of interpreters to simply deny the real existence of one of the two minds, namely, the inadequate one corresponding to subjective experience.

8.2 The Intellectualist Reading of the Mind

Starting with Hegel, who in his *Lectures on the History of Philosophy* presented Spinoza as the thinker of absolute substance, crushing individual subjectivity under its totality (Hegel 1974, 252–90), intellectualist views of Spinoza have been legion throughout the history of ideas. One of these, leading to the thesis that subjective experience has no ontological room in Spinoza's system, was famously expressed by one of the first important Spinoza scholars of the early twentieth century, Harold H. Joachim. His words couldn't be stronger:

> There is a fatal trend in Spinoza's philosophy toward abstraction, in spite of all his struggles towards the conception of a concrete unity. Thus, things in their temporal being—the actual world of the perceptive consciousness—either turn into illusions, or slip back into the world of eternal timeless necessity, the universe of science . . . Our actual mind with its emotions, volitions, desires, is *qua passional* unreal. In its reality it is a part of the 'infinita idea Dei'; but in the completeness of that 'idea' all passion vanishes. (Joachim 1901, 96)

Joachim heavily criticizes Spinoza for this.[10] He does see that Spinoza strives to avoid reducing the subject to nothing, but his interpretation does not manage to go beyond the contradiction between the unity of substance and its different attributes and modes, on the one hand (an ontological problem), and between the absoluteness of God's knowledge and the finiteness of our subjective experience, on the other hand (and epistemological problem). When explaining this contradiction, he speaks of a "conflict" in Spinoza's theory, a view that is "not consistent," and he clearly says that in Spinoza subjective experience is "not brought into harmony with his general principles," that it is "irreconcilable," and that it "comes into positive collision" with them:

> The modal apprehension is in part illusory, and the illusion is a fact—and yet a fact for which no place can be found in Spinoza's conception of the ultimate nature of things. He

[10] This discussion is primarily found in the Appendix immediately following, entitled "Difficulties and criticisms" (Jaquet 2005, 98–122).

> describes the fact in terms of his general theory, but his description is no explanation; and *if taken as an explanation* it conflicts with his statement of the general nature of God ... An illusion must fall somewhere; for Spinoza, therefore, it must 'be' in God. And the question is how this is possible. It is no answer to this question to say that it is in God in so far as God is himself the product of an illusory apprehension, and yet 'God as affected *in infinitum* by infinite modifications' is not consistent with God as the 'absolutely complete positive being' which—Spinoza has shown us—an ultimate apprehension demands. It seems clear, then, that the world of presentation and 'natura naturata' as an order of distinct modes are in some sense 'facts,' which Spinoza has not brought into harmony with his general principles ... His conception of the infinity of completeness is irreconcilable with the indefinite infinity of the finite. (Joachim 1901, 112–13)

This is an interpretive problem that no one can ignore. Joachim traces it back in the first place to Spinoza's geometrical method, requiring that the *Ethics* express the point of view of absolute knowledge, so that error can only be "a privation of knowledge," as E IIP35 holds,[11] while a privation can be nothing ontologically. For him, there is simply no answer to be found as long as we remain in this absolute perspective, because it will always necessarily entail that concrete and subjective singularity be negated.

> If my interpretation of the immediate and mediate infinite and eternal modes is correct[12]—and I am unable to see my way to a better one—all the distinctive features of the worlds of Extension and Thought seem to vanish as 'illusions' one by one, until you are left with the singleness of the Attributes: *a singleness not concrete, but abstract*. Spinoza is indeed far too ready to dismiss things as 'mere illusions'. (Joachim 1901, 114)

Is there a way to escape this fatal conclusion that Spinoza simply cannot account in positive terms for what is not an absolute being and/or knowledge[13] in God? Joachim's remark is absolutely true: even "an illusion must fall somewhere" (Joachim 1901, 113, quoted above); in other words, there *must* logically be some place for it "in" God. But is Joachim equally right in claiming that "It is no answer to this question to say that it is in God only insofar as God is himself ... 'affected

---

[11] E IIP35; G II, 116: "Falsity consists in the privation of knowledge which inadequate, or mutilated and confused, ideas involve." We shall explain this in our third section.

[12] Joachim's reading of the immediate infinite modes is absolutely standard (Jaquet 2005, 82 *sq.* for the attribute of extension, and 93 *sq.* for the attribute of thought). As far as the infinite mediate mode in the attribute of thought is concerned—which, contrary to the others, Spinoza does not specify in Ep. 64 to Schuller (G IV, 278)—Joachim, like many later commentators, takes it to be the infinite "idea of God." In the past, I have personally sided with Beyssade in considering that the infinite *idea Dei* is actually equivalent to God's infinite intellect, on the basis that an idea is an act of understanding, and I have instead proposed to follow his suggestion that God's infinite *acquiescientia in se ipso* and love of himself are the unidentified "infinite *mediate* mode", since they follow of all eternity from God's self-knowledge [see S. Malinowski-Charles (2004, 142–3) and Beyssade (1994)]. However, I am now tempted by the other alternative, because if we have an "idea Dei" which is different from the infinite intellect, we can see it as encompassing absolutely all ideas, even the inadequate ones (whereas the intellect can only know in a true and adequate way). In any case, the problem that Joachim mentions in this quotation remains the same even with the other interpretation of the infinite mediate mode in thought.

[13] According to E IIP7 and C (G II, 89), being and knowledge go together since an idea must objectively correspond to any single thing in God.

in infinitum by finite modifications'?" (Joachim 1901). This is, obviously, where the whole difficulty resides. Contrary to this reading of Spinoza, my aim in this paper is to show that the finite and truncated idea that constitutes our mind is also truly, *even in its finiteness and inadequacy*, a reality within God which somehow overlaps the absolute reality corresponding to us from an absolute viewpoint. But the time has not arrived yet for such a demonstration. For now, I shall continue to show how widespread, and surely much more than is consciously acknowledged, this idealist tendency is in English interpretations of Spinoza.

The tradition negating the reality of subjective experience, of which Joachim was one of the leading spokesmen, bears a certain relation to the "subjectivist reading of the attributes" that was also quite fashionable in the early years of Spinoza scholarship. According to this view—which was held, in particular, by Constantin Brunner (1968, 442–448; 1976),[14] but also before him by J. E. Erdmann (1834, I, 2, 60 *sq.*) (an Hegelian) and Pollock (1880, 175–179)— (who wrote the first biography of Spinoza in English) thought and extension (among other attributes) are just illusory ways of introducing differentiation within the absolute substance. In other terms, Spinoza's monism would mean something like Parmenides' unmoving and unchanging "One"—at least how it has usually been conceived in the metaphysical tradition—i.e., a being in which any seeming "parts" are just phenomena related to individual perception, and are not rooted in any ontology whatsoever. The comparison is mine, but it does express adequately, I believe, the main line of this reading. Relying on Spinoza's definition of the attribute as "what the *intellect perceives* of a substance as constituting its essence" (E IDef4; G II, 45), these authors consider that the true unity of mind and body can only be accounted for if we cease to consider them as two *different* modes, and rather focus exclusively on the fact that "the mind and the body are one and the same thing, which is conceived now under the attribute of thought, now under the attribute of extension" (E IIIP2S; G II, 141). This view, taken to the letter, and actually beyond, leads these authors to an elimination of any real distinction between the attributes within the substance.

Needless to say, this idealism agrees with, and actually reinforces, the conclusions presented above by Joachim about the illusoriness of subjective experience. The subjectivist reading of the attributes introduces a gulf between reality, which is one, and consciousness, which is manifold. These two forms of rationalist reading thus concur in rendering unintelligible such things as subjective experience, finiteness, and change at the modal level. The whole modal realm of finite things, precisely because of its finiteness, progressively vanishes into nothingness.

Is it because of Martial Gueroult's excellent refutation of the subjectivist reading of the attributes[15] that commentators in the French tradition have not followed this same line of interpretation?[16] Once Gueroult showed that this reduction was wrong

---

[14] These are posthumous re-editions.

[15] See Gueroult (1968, 428–61), vol. I, App. 3: "La controverse sur l'attribut."

[16] To my knowledge, no one in the French tradition has taken up such views, at least in the last 40 or 50 years.

and unfaithful to Spinoza, it also followed that any attempts in the same style were cut at the root. But such was not the case in the English tradition (which may also have been more Hegelian than the French one), and this kind of "absolutist", and absolutely rationalist reading of Spinoza was perpetuated.

One may think that it is only an old story, and that the question as to the ontological place of subjective experience, and as to what constitutes our "self", is no longer ours to ask. But a line of influence can be traced, extending, in broad outline (despite individual variations that I do not mean to reduce, although I cannot do them full justice in the frame of this paper), from Joachim to Wolfson (who explicitly defended a subjectivist reading of the attributes [1934, vol. I, 142–57; and more particularly, 145–7]), and through him and possibly others, including Curley's interpretation of the attribute of thought as a logical set of propositions leaving little space to subjectivity,[17] to Bennett (1984), Scruton (1986), and to the most contemporary of our contemporaries.

This assimilation of different names may need nuancing. Indeed, Bennett does not say anything similar to what Joachim says about the unreality of our subjective (and passive) experience. But he does posit the problem of the two "I"s or minds by arguing that Spinoza had a "tendency to conflate logic with psychology" (1984, 52), and by saying that in fact "[t]here is a way of taking his term 'thought' which divorces it from psychology" (1984, 50). The problem of the two minds is treated differently, but it is clear that Bennett notices the difficulty in Spinoza, and rests on an acknowledgement of failure on his part to give mental experience ("psychology") a full status within his metaphysical doctrine: "from time to time he makes his psychology double as logic as well, taking the term 'idea' to stand indifferently for a mental item and for a concept or proposition" (1984, 52).

Scruton, as far as he is concerned, fits even more clearly with the idealist reading. His conclusion on the 'problem of the two minds', as I have dubbed it, is perfectly unambiguous:

> Spinoza's monism generates a highly paradoxical idea of the human person. *The individual person is not, it seems, an individual at all.* Nor is anything else. The identity, separateness, and self-sufficiency of the person *all seem to be denied by Spinoza*, and man, as part of nature, seems to be no more important a feature in the scheme of things than the rocks and stones and trees. (Scruton 1986, 53)[18]

And finally, in a recent paper entitled "Rationalism Run Amok: Representation and the Reality of Emotions in Spinoza," Michael Della Rocca refers approvingly

---

[17]It should be noted that Curley argues both against Joachim, and against Wolfson, whom he (rightly) sees as mutually opposed on the question of the reality of the modes, before moving to his own interpretation meant to reconcile them (1969, 20–36). I should also mention that Curley does reintroduce psychology in Spinoza's attribute of thought, but only at the human level, because he sees it as exclusively endowed with consciousness or "idea of idea."

[18]Italics added. Genevieve Lloyd, who explicitly refers to this passage at the beginning of her book (1994, 6), may well have chosen its title (*Part of Nature*) as an answer to this view she criticizes, if not in reference to a phrase by Spinoza in Ep. 32 to Oldenburg (quoted below). She also attacks Scruton on p. 43.

to Joachim[19] as a forerunner of the very view he himself defends, namely, the conception that I am surely something in God, but that this I is not "mine."

> Insofar as my mind is active it is *in* God's intellect [cf. E 5P40S, quoted before]. The suggestion is that, by contrast, the mind—insofar as it is passive—does not help to constitute God's intellect and so is not *in* God. And this is to say that my mind qua passive is not fully intelligible and does not fully exist. (Della Rocca 2008a, 50–1)

In this paper, he argues that there is less reality to my passions and inadequate ideas than to my adequate ideas. Again, Della Rocca's arguments rely on a hyper-rationalist view of Spinoza, also expressed in his latest book (2008b), according to which Spinoza is led from the beginning to the end of the *Ethics* by the Principle of Sufficient Reason (PSR) requesting a (rational) reason for everything in the world, and for every claim he makes. As Della Rocca says in a way of (ironical) conclusion,

> This [Spinoza's fundamental charge against the affects] is simply an instance of the more general insight that passivity is not fully real, that passivity strips things of their existence to some degree. This charge is propelled by Spinoza's PSR.... Thus in Spinoza's eyes, when it comes to the affects—as in so many other things—all positive and negative metaphysical judgments are dictated by the PSR. For Spinoza, the PSR giveth and the PSR taketh away. (Della Rocca 2008a, 52)

I hope to offer a more "empiricist" reading of Spinoza in what follows. In doing so, I will make the case that Spinoza's account of subjective experience is neither contradictory, nor even problematic in light of his theoretical premises.

8.3 That Inadequate Ideas Are Also in God

What is hard to understand, as we have seen, is the status of my inadequate or imaginative ideas. Are they in God too? The interpretations discussed above all end up with the reductionist view because they maintain that these ideas, if they really existed, could only be in God's *intellect*, which contradicts the view that the intellect is always characterized as perceiving things truly. In what follows, I will attempt to show that instead of negating one of the two minds, we achieve a much more satisfactory (and common-sense) account of Spinoza's views if we strive to understand how something both non-eternal, and inadequate, can *also* be given in God. In a nutshell, my contention is the following: the idealists are right in saying that "my mind" in God's intellect is not the one I experience; but they are wrong in assuming that this mind is the only one which is "in" God. I even take it to be the ultimate meaning of God's "modifiability" that there are two minds superimposed one onto the other, and that God (or nature, or substance), as the all-encompassing reality, takes each and every form that duration or the modal (transitive) line of causation brings to existence. In other words: I am in God in two ways, one temporal, the other eternal.

---

[19] See for instance his quotes of Joachim on pages 48 and 50.

First of all, we must understand why our mind, insofar as it is finite, is necessarily an inadequate idea, and why this inadequacy is described in terms of a "lack" or "privation" of knowledge by Spinoza ("Falsity consists in the privation of knowledge which inadequate, or mutilated and confused, ideas involve," E IIP35; G II, 116). This explanation is mainly found in the last part of E IIP11C, of which I only quoted the beginning when, in the first section, I presented the passages that opened the way for the intellectualist views. Interestingly, there is also an answer to their problems in the last part of the corollary. Quoted in full, this corollary states the following:

> From this it follows that the human mind is a part of the infinite intellect of God. Therefore, when we say that the human mind perceives this or that, we are saying nothing but that God, not insofar as he is infinite, but insofar as he is explained through the nature of the human mind, *or* insofar as he constitutes the essence of the human mind, has this or that idea; *and when we say that God has this or that idea, not only insofar as he constitutes the essence of the human mind, but insofar as he also has the idea of another thing together with the human mind, then we say that the human mind perceives the thing only partially, or inadequately*. (G II, 94–95; italics added)

The truncation of our ideas is a metaphysical necessity related to our situation not as "an empire within an empire,"[20] but as a *part* of nature.[21] As an idea of a particular body, our mind only grasps the other bodies through the ways in which they affect "its" body, and not in themselves. My mind's perception is thus necessarily inadequate with respect to the other bodies in the world (even if it must also be true with respect to my body).[22] Henry Allison summarizes Spinoza's argument in a very clear formula:

> Perceptual ideas reflect the condition of the organism in its interplay with the environment (which is their actual "object," or correlate), rather than the true nature of some independent reality. Thus, insofar as the mind takes such ideas to represent some external thing as it is in itself, rather than the manner in which that thing affects its own sensory apparatus, it inevitably falls into error. (Allison 1987, 107)

To continue with E IIP11C, the part that I have italicized reveals something else of great interest, namely, that this partial or inadequate cognition is what we call the "human mind" only when "God has this or that idea . . . insofar as he also

[20]This is a phrase famously used by Spinoza in the Preface to *Ethics* III (G II, 137).

[21]One may find an excellent account of error and inadequacy, in perfect line with the realist view of the modes and of knowledge that I defend here, in Lloyd (1994, 43–73), section II: "Knowledge, Truth, and Error." I am very indebted to Lloyd for the clarity of her explanations and the rightness of her philosophical intuitions. Her chapter is much better than my own explanation of the same subject (2004, 147–59).

[22]It is true insofar as it is perfectly corresponding to my body's state, as Spinoza naturally deduces by saying that "whatever happens in the object of the idea constituting the human mind must be perceived by the human mind" (E IIP12; G II, 95), a statement rephrased in an even more precise way in the corollary to E IIP13: "From this it follows that man consists of a mind and a body, and that the human body exists as we are aware of it" (G II, 96). The point that the finite mind is necessarily both inadequate and adequate, with respect to the other bodies or its own, can be found in Bartuschat (1994, 187 and 200–3).

has the idea of another thing together with the human mind" [*quatenus simul cum mente humana alterius rei etiam habet ideam*]. The "also" here (*etiam*) confirms my reading that there are really two minds, both of which are given in God (i.e., that are both modifications of him).[23] In other terms, the human mind *is* God, or God *is* the human mind, but under certain specific conditions. These conditions, I believe, are the conditions of *modal experience*.

This inference can be made by comparing this corollary to another difficult, but highly informative, passage:

> E IIP9: The idea of a singular thing which actually exists has God for a cause not insofar as he is infinite, but insofar as he is considered to be affected by another idea of a singular thing which actually exists; and of this idea God is also the cause, insofar as he is affected by another third, and so on, to infinity.
>
> Dem: The idea of a singular thing which actually exists is a singular mode of thinking, and distinct from the others (by P8C and S), and so (by P6) has God for a cause only insofar as he is a thinking thing. But not (by IP28) insofar as he is a thinking thing absolutely; rather insofar as he is considered to be affected by another mode of thinking. And God is also the cause of this mode, insofar as he is affected by another, and so on, to infinity. . . .
>
> Cor: Whatever happens in the singular object of any idea, there is an idea of it in God (by P3), not insofar as he is infinite, but insofar as he is considered to be affected by another idea of singular thing (by P9); but the order and connection of ideas (by P7) is the same as the order and connection of things; therefore, knowledge of what happens in a singular object will be in God only insofar as he has the idea of the same object, q.e.d. (G II, 91–2)

There are many doctrines to be derived from E IIP9 and what follows, but the main one for our concerns is surely that, again, God *is also* the finite mind, yet he is this human mind only insofar as he is "affected by another mode of thinking." The demonstration here refers explicitly to E IP28, in which Spinoza, after having presented the immanent causality in God as a deduction from substance (or attributes) to infinite and then finite modes, comes to present the transitive causality which holds between the finite modes in the frame of the infinite modes that they constitute and characterize:

> Every singular thing, or any thing which is finite and has a determinate existence, can neither exist nor be determined to produce an effect unless it is determined to exist and produce an effect by another cause, which is also finite and has a determinate existence; and again, this cause also can neither exist nor be determined to produce an effect unless it is determined to exist and produce an effect by another, which is also finite and has a determinate existence, and so on, to infinity. (E IP28; G II, 69)

It is then to a "horizontal" chain of causality of one finite thing onto another that *we*, as finite things, belong, and this (mechanical) chain of causality is also infinite, though in a different sense than the infinity of the substance-to-mode causation (which, obviously, is ultimately a relationship of identity). Our mind is God insofar

[23] Another little word that offers a strong textual justification for this dual view of the mind is the "*primum*" that Spinoza uses when he says that "The *first* thing which constitutes the actual being of a human mind is nothing but the idea of a singular thing which actually exists" (E IIP11; G II, 94).

as it is affected by other ideas, just like our body is God insofar as it is affected by other bodies: to say it shortly, insofar as it is *a finite, determinate mode*.[24]

The modal status is the one in which experience and consciousness can be given. It is *as* a finite mind, and *as* a finite body, that God or substance is (in) me and that I am (in) him. In this respect, it seems to me undeniable that Spinoza does locate inadequacy within God. But again, it is simply not within his infinite intellect, for as we have seen, "all ideas, insofar as they are related to God, are true" (E IIP32; G II, 116), which is to say that the idea which God *has* of my body is necessarily true. The idea which he *is*, and which I, as a finite mode of him, also am and may experience (if I am sufficiently complex to have conscious sensations[25]), is necessarily inadequate because it is the perception of a mere part of an infinite chain of modal causality. This is, to reiterate, what E IIP11C said: "When we say that God has this or that idea, not only insofar as he constitutes the essence of the human mind, but insofar as he also has the idea of another thing together with the human mind, then we say that the human mind perceives the thing only partially, or inadequately."

There are many other passages in Spinoza that offer arguments for proving that experience and inadequate ideas are in God, and are not mere "nothings". E IIP3D, for instance, explains the proposition that "[i]n God there is necessarily an idea, both of his essence, and of everything which necessarily follows from his essence" (E IIP3; G II, 87) by saying the following:

> For God (by P1) can think infinitely many things in infinitely many modes, or (what is the same, by IP16) can form the idea of his essence and of all the things which necessarily follow from it. But whatever is in God's power necessarily exists (by IP35); therefore, there is necessarily such an idea, and (by IP15) it is only in God, q.e.d. (E IIP3D; G II, 87)

Perhaps interpreters in the intellectualist trend will object that Spinoza speaks here of "all the things which *necessarily* follow from him," by which they will say that only the *immediate* infinite mode, i.e., the *intellectus infinitus Dei*, really follows necessarily. I reject this latter assumption, since necessity and determination are for Spinoza everywhere in nature, and the "horizontal" modal determination is no less a necessity than the "vertical" immanent deduction from the attributes. Furthermore, Spinoza's appeal to E IP35 in the proof is hard to reconcile with the idea that Spinoza does not talk about all the finite modes, but only about the infinite intellect.

Interestingly, some of the arguments for the realist reading of inadequacy that I defend here can be found (as in the case of E IIP11C) in the very passages that led

---

[24]Both for an explanation of how the two chains of causality (vertical and horizontal) are mutually constitutive, and of the importance of the adverb "*quatenus*" ("insofar") to understand Spinoza's monism, see my first chapter, "L'unité causale dans le tout de la nature," in my (2004 21–36).

[25]Spinoza links both the degrees of "animation" of the singular beings and their capacity for "clarity and distinctness" with the complexity of their body in E IIP13S (G II, 97). On this, see footnote 4 in section I.

the supporters of the hyper-rationalist interpretation to the conclusion that subjective experience, imagination, and the passions deriving from this kind of knowledge, are unreal illusions or, at best, lesser forms of reality. This is so because they rest on the position that an idea can only be given in God's *intellect*, and as such must be adequate, so that anything that cannot fit there is seen by them as fitting nowhere.

One of these controversial passages is E IIP33, whose demonstration is particularly interesting:

> E IIP33: There is nothing positive in ideas on account of which they are called false.
>
> Dem: If you deny this, conceive (if possible) a positive mode of thinking which constitutes the form of error, or falsity. This mode of thinking cannot be in God (by P32). But it also can neither be nor be conceived outside God (by IP15). And so there can be nothing positive in ideas on account of which they are called false, q.e.d. (G II, 116)

Spinoza raises for himself the problem that we are confronted with, i.e., that of locating where inadequate ideas may go, and his answer is indeed that there is nothing *in ideas themselves* that constitutes the "form" of error. This is easy to interpret as saying that inadequate ideas are *themselves* nothing, and are not in God. However, I think that we can make sense of this claim in a very different way, and that this reading couldn't be more wrong or hasty. For the same demonstration also very clearly states that if there is a mode of thinking, then it must necessarily be in God, an assumption made with reference to E IP15 (according to which "Whatever is, is in God, and nothing can be nor be conceived without God"; G II, 56). It would be nonsense to believe that Spinoza's aim in this demonstration is to say that inadequate ideas are not modes of thinking. The proof goes the other way round: since they *are* modes of thinking, as experience testifies, then we must understand that they are wholly positive in themselves (and that their "error" is extrinsic to them).

In an article that, despite what its title suggests, only partially deals with the problem of the two minds at stake here, Bartuschat (1994) interestingly remarks that it is experience that gives the system its content, and not the opposite. By this, he means that

> *Ethics I* supplies only a structural analysis of God, with respect to the concepts of substance, attribute and infinite and finite modes. It does not fill these terms with a determined content. Spinoza lists their contents as empirical givens and does not deduce them from God. (Bartuschat 1994, 189)

Indeed, it is in axioms (that is, in non-provable facts of experience) that Spinoza introduces thought and extension, namely through E IIAx2: "Man thinks" (G II, 85), and E IIAx4: "We feel that a certain body is affected in many ways" (G II, 86). As Bartuschat further argues,

> It is only after Spinoza's axiomatic introduction of human thought and human corporeity in *Ethics* II that the contents of God's attributes are determined as thought and extension ...

> Although thought and bodies cannot be deduced from God, who neither thinks nor is extended;[26] *if they exist—and there is no doubt that they do, since man thinks and feels his body—they have the character of modes.* (Bartuschat 1994, 190; italics added)

Far from leaving no room for subjective experience, then, we can in fact read Spinoza as a philosopher *of* experience, whose rationalist and systematic framework only serves the purpose of accounting for experience, that is, of coming to its service in order to make it reach new dimensions (those, it is true, of a rational conduct). So, again, whatever we feel as existing – be it our ideas or emotions, even the passive ones – must also be attributed to God.

Let us sum up what has been shown thus far. Spinoza's explanation of the inadequacy of our ideas in the first kind of knowledge, i.e., in "imagination," is given in the second part of the *Ethics,* between propositions 11 and 37. There, Spinoza starts from the idea that a mind must necessarily be the mental counterpart to the body that we have in the attribute of extension, and then he moves on to the idea that this whole, or complete, mind, is *not* the one that we experience through our self-awareness—this is what was shown in section I of this paper. There must be, then, two distinct ideas corresponding to what we could call my "mind": one adequate or complete, corresponding necessarily in God's infinite intellect, in virtue of the parallelism of the attributes, to my body as it can be explained through the laws of nature (and this idea, as we have seen, is also necessarily abstract), and one inadequate or incomplete, corresponding to my body as I experience it. The subjective mind is also in God; in fact it *is* God, but as a finite mode. In addition, the mind that I experience must somehow be a *part* of my whole mind, precisely because all that it is "is" something, which Spinoza puts as follows: "There is nothing positive in ideas on account of which they are called false" (E IIP35). What still demands explanation is just the sense in which the incompleteness that makes the first kind of cognition represent things "in a way which is mutilated, confused, and without order for the intellect" (E IIP40S2; G II, 122), is both nothing (with respect to God's infinite intellect), and something (with respect to modal experience). The question that arises here is: How can we understand experience as grounded in logic or abstraction? That is, in what sense is the subjective mind a "part" of the purely objective mind?

---

[26] This phrase relates to a very odd part of his interpretation, argued at 193–5 and repeated later, according to which "thinking is a modal determination" (194), i.e., the *attribute* of thought does not itself produce ("cause") the ideas as objective essences of the things in the other attributes, whereas the other attributes themselves cause their modifications. On p. 198, again, he reiterates that "objectivity is therefore not grounded in God himself, but in a mode: the infinite intellect." I confess that I cannot make sense of this claim, except maybe as an attempt to eliminate potentiality from God, because Bartuschat also insists that finite thoughts must be produced together with, i.e., at the same time as, the formal modifications in the other attributes. His point is thus that all thoughts can only be effects. But this leads him into a dangerous process of correction of the passages in which Spinoza presumably (in his view) meant something other than what he explicitly wrote (see 193–4). I do not follow him here.

8.4 How the Order of Imagination is "Superimposed" onto the Order of the Intellect

In a letter written by Spinoza to Oldenburg on November 20, 1665, i.e. at a date when his metaphysical system was now definitely set, Spinoza addresses exactly the question of this part-taking of the finite, human mind within the infinite intellect of God:

> You see, therefore, how and why I think that the human body is a part of nature. But as far as the human mind is concerned, I think it is a part of nature too. For I maintain that there is also in nature an infinite power of thinking, which, insofar as it is infinite, contains in itself objectively the whole of nature, and whose thoughts proceed in the same way as nature itself, its object, does. Next, I maintain that the human mind is this same power, not insofar as it is infinite and perceives the whole of nature, but insofar as it is finite and perceives only the human body. For this reason I maintain that the human mind is a part of a certain infinite intellect [*hac ratione mentem humanam partem cujusdam infiniti intellectus statuo*]. (Ep. 32; G IV, 173–4; italics added)

No one will deny that our bodily affections *are* something, so perhaps it will be easier to start with this image in order to explain that "the human mind is [the] same power as the infinite power of thinking," but simply "insofar as it is finite and perceives only the human body." Let us say that my face blushes when I am intimidated or when I get angry. According to the parallelism of the attributes, there are inadequate ideas corresponding to this intimidation or anger that are given in my mind, just in the same way as the blushing and other physical manifestations (say, an increase in my heartbeat) are given in my body. If it were true that these inadequate ideas are nothing existing, then we would have to say that the blushing and the augmentation of my cardiac pace are sheer illusions too, which is absurd. Rather, we can understand from this example that as long as a bodily state is given, there is an idea to express it. But this bodily state itself can be understood in two ways: one, sensitive or experiential, is given in the subjective mind; the other, rational, is given in the objective mind that perfectly corresponds to my body in God's infinite intellect. That there is blushing and a certain heartbeat in my body is a fact: these things are bodily affections. And since anger or intimidation are not just affects for the mind, but actual forms of transition of my body from a higher perfection to a lower one (Spinoza places both these affects in the category of sadnesses or pains in *Ethics* III), we must also say that these bodily affections are affects. (It may be useful to recall here that the affects are "the affections of the body by which the body's power of acting is increased or diminished, aided or restrained, and at the same time, the ideas of these affections" [E IIIDef3; G II, 139].)

God will, surely, have an idea of these affections as facts. But does God "feel" them as affects? It looks like we are back at the problem raised by Bayle in his *Historical and Critical Dictionary*, where he ironically criticized Spinoza for his supposed view that "God, modified as ten thousand Germans, killed God, modified as ten thousand Turks."[27] Malebranche formulated a similar argument and used it in

[27] Bayle (1740, 261), vol. IV, Article 'Spinosa,' Note N. No. IV.261.

the ninth of his *Dialogues on Metaphysics and on Religion* (first published in 1688), namely in a section devoted to the problem of evil, in order to justify his view that pantheism is untenable. His words are even more eloquent:

> Even the author who has revived this impiety agrees that God is the infinitely perfect Being. And that being so, how could we believe that all the created beings are but parts or modifications of the divinity? Is it a perfection to be unjust in one's parts, unhappy in one's modifications, ignorant, demented, impious? There are more sinners than good people, more idolaters than believers. What disorder, what discord between the divinity and its parts! What a monster, Aristes, what an appalling chimera! A God necessarily hated, blasphemed, scorned, or at least unknown by the better part of what He is. For how few people would think of recognizing such a divinity? A God who is necessarily either unhappy or unfeeling in most of His parts or modifications, a God who punishes or exacts vengeance on Himself. In a word, an infinitely perfect being nonetheless comprising all the disorders of the universe. (Malebranche 1997, 150)

To both of these views claiming that god feels all the pains affecting its modes, my answer would be: "Yes, and no." Yet again, everything relies on the "insofar" of the answer. Yes, this is true of Spinoza insofar as God is considered as modified by an infinite chain of (horizontal) finite causes. But no, this is not true—at least, as far as the *affective* part is concerned—when we look at God's infinite essence and its immediate mode, the infinite intellect. It remains true though, even at this level, that "God, modified as ten thousand Germans, killed God, modified as ten thousand Turks," but it is true as a *fact*, or as an *affection* (a modification) of God's infinite power not as an affect at this level. The conclusion then needs to be twofold: it appears necessary that in God's infinite intellect there be an idea of all the states of his modes, i.e., of all his "affections," but the *affects* are in God only insofar as he is modified as a finite human being (or some other kind of being having affects). For as a totality, it cannot express any change. As Spinoza relates to Schuller in his Ep. 64 concerning the infinite mediate mode in the attribute of extension, "the face of the whole universe [*facies totius universi*],. . . although it varies in infinite ways, nevertheless always remains the same" (Ep. 64, G IV, 278).

To the example given above of the blushing that results from my intimidation or my anger, God as finite (that is, me) does perceive and feel these affective states; but as an infinite being, it only contains the ideas (and, in the attribute of extension, the actual bodies) that explain why this blushing mechanically happened from the previous state of my body: the blood's motion, being made quicker because of certain factors, became salient on my cheeks because of certain other factors, etc. We find here again the abstract level of thought that we defined as comprised by general laws of logic, mathematics, etc. Ultimately, everything "happening" (all the *facts* or *affections*) can be reduced in terms of a given ratio of motion and rest, and God's intellect links the affections with their true, objective causes. Absorbed as I am in my emotion, I do not have any consciousness of what actually explains both the emotion itself and the new state into which it propels me, and which is "my mind" as objective adequate idea of my body. My consciousness, which is related to my experience, is of course a thought in God too, but it is, to use the vocabulary I have already introduced, a thought that God *is* rather than that he *has*. Or rather, he *is* and rather than that *has* both ideas, but in different senses.

This issue is extremely difficult, and this interpretation may or may not be the good one. But wagering that it is, I would like to use another image to clarify how inadequate ideas are given in addition to adequate ideas, as well as what it is that constitutes this difference between inadequate ideas as facts and as experienced affects.

Imagine a wood surface into which a child hammers some nails. The nail heads may be disposed in such a way as to form a particular figure or picture. Linking them with threads will make this figure appear out of what was initially just a random, confused disposition of dots on the surface. Let us presume that the clear figure is the adequate understanding of the dots on the tablet, i.e., is the adequate link between them. Let us say, now, that a child sees the dots on the surface and links them inadequately. The figure he will make with his thread will not be the "right" or "true," one. It may be interesting indeed, and it may even form a picture, but that picture will not be the one corresponding to the objective link of these nail heads as they were originally ordered. Will his picture express *nothing*? Surely not: it will reflect the mechanical order followed by the child and leading from one dot to another.

Now, in this analogy,[28] the nail heads or dots correspond to the particular determinations that compose my mind and my body (affections in the body, perceptions or ideas in the mind), and the thread is knowledge and consciousness. Adequate knowledge is the perfect picture of the order and concatenation of these determinations, as they can be explained by eternal and universal laws, whereas inadequate knowledge makes links between these felt events and ideas that do not respect the right causal order explaining their concatenation. What I think this example makes clear is that the dots on the panel themselves (i.e., the affections or the ideas) are identical in the two cases. It is only the way of linking them which is different, and which in one way is clear, in the other, confused.

To apply this image to Spinoza's theory of the mind, the dots or affections are "me" in both cases. Furthermore, all that there is, is something (either an affection, or a conscious knowledge). Where is error or falsity in this schema, then? Nowhere. Falsity is only a judgment that can be made by comparing the two figures traced; it is neither of them *in themselves*. It is probably this that Spinoza has in mind when he writes that "there is nothing positive in ideas on account of which they are called false" (E IIP33, quoted above). The subjective mind ("my mind") has ideas that are "truncated" in the sense that the threads linking the dots *miss* several chunks of the right causal chain corresponding to adequate knowledge. Falsity is nothing in itself, and can only be spoken of if we compare my ideas to those in the objective mind, where each affection is linked to its adequate cause according to the eternal laws of

---

[28]This analogy is faulty insofar as it introduces a temporal discrepancy between the "original pattern" and the one traced by the child. To have a better comparison, we would need an example where the two orders are simultaneous.

the intellect. This was also the argument of the scholium to E IIP17, where Spinoza first defined imagination as a positive power and expanded upon the question of error:

> Here, in order to begin to indicate what error is, I should like you to note that the imaginations of the mind, considered in themselves, contain no error, or that the mind does not err from the fact that it imagines, but only insofar as it is considered to lack an idea which excludes the existence of those things which it imagines to be present to it. For if the mind, while it imagined nonexistent things as present to it, at the same time knew that those things did not exist, it would, of course, attribute this power of imagining to a virtue of its nature, not to a vice. (G IIP17S; G II, 106)

I can see two main reasons why knowledge is necessarily different in me subjectively than objectively in God.[29] Firstly, the chain of finite causes extends to infinity, whereas my point of view is restricted to my body (I only know the other ones through the way in which they affect me).[30] Secondly, I link these dots by following the order of their succession and/or the causality I *imagine* between them, rather than in the causal order that explains them objectively. It is a sort of order too, but not the right one. This idea is reminiscent of what Spinoza says in E VP10, when he contrasts the random ordering of our affections in the first kind of knowledge with "the power that I have of ordering and connecting the affections of the body according to the order of the intellect." In the scholium to this proposition, he asserts:

> By this power of rightly ordering and connecting the affections of the body, we can bring it about that we are not easily affected with evil affects. For (by P7) a greater force is required

[29]Bartuschat (1994) makes a similar claim in a very literal sense, since he takes this necessary difference as something impossible to overcome. Namely, he insists on the fact that what is given in the subjective mind cannot possibly be what is in objective knowledge, and that conversely, the infinite idea of God cannot possibly include the finite subjective view: "It is only by imagination that man knows his body, which exists in temporal extension ... While man knows his body only via the ideas of its affections, the idea of the body is in God insofar as God constitutes not only the human mind. Therefore, this idea is nothing for man ... As such, this idea is something indefinite that cannot be experienced by human beings" (202–3). And conversely: "The infinite idea, which is in God, does not have the whole range of temporal changes of its objects" (203). Is this really so extreme? Can the two minds not "communicate" more; i.e., is there really no bridge—no identity—between them? It is hard to take a position on this question, and Bartuschat definitely agrees with us in the distinction he is making between the two minds that correspond to me. Opposing his view on this would mean introducing the finite point of view in God's intellect or idea, and saying that the inadequate consciousness which is mine is contained in it. On the one hand, I find this reading (which is far weaker than that of Bartuschat) quite plausible, to the extent that it takes seriously the full meaning of the idea that "I" am part of God's infinite intellect. On the other hand, I am very sensitive to his argument that this is impossible, since the infinite intellect contains all ideas, and thus "the idea of the body is in God insofar as God constitutes not only the human mind" (but also all minds). I would even see another argument for it at E IIP43D: "An idea true in us is that which is adequate in God insofar as he is explained through the nature of the human mind (by P11C)" (G II, 123).

[30]See E IIP26; G II, 112: "The human mind does not perceive any external body as actually existing, except through the ideas of the affections of its own body."

> for restraining affects ordered and connected according to the order of the intellect than for restraining those which are uncertain and random. (E VP10S; G II, 287)

Simply said, the ordering of the determinations constituted by my experience can be random (if my knowledge of them is inadequate), or it can fit the order of the intellect (if I link adequately the effects with their right causes). My experience corresponds primarily to what Spinoza refers to as the "order of nature" or "of encounters," as contrasted with "the order of the intellect." But the former is grounded in the latter, in so far as the order of the encounters that links (and produces) the dots in a given way is also explained by the laws according to which the adequate knowledge of their chain arises. As E IIP36 states, "[i]nadequate or confused ideas follow with the same necessity as adequate, or clear and distinct ideas" (G II, 117). The inadequate links I make between the determinations of my body are still in God somehow, since they are features of thought, i.e., ways of thinking, which means that God's thought does not take the form of his infinite intellect *only*. Surely, there is no "infinite imagination", but this is so because imagination is *always* related to finiteness. There are conditions for this imagination to happen, and these conditions are empirical: they are those of any finite mode (i.e., those of duration, and of passivity related to its environment, of any finite existing thing), and additionally, they are those of a minimal bodily and mental complexity.[31] In the case of humans, it is clear that both of these conditions obtain.

Thus, we have two minds indeed, and the intellectualist commentators are wrong in believing that the necessarily logical aspect of the attribute of thought in Spinoza precludes the possibility for a psychology to be *also* given by him – this is, in particular, a claim that Bennett had made. These two minds are related to each other insofar as the objective being that we are in God's intellect explains, and justifies, the subjective being that we are in God as finite. The fact that one is experiential, and the other one logical, is non-problematic for their mutual relation. But the experiential mind only exists insofar, and as long as, its particular object exists, i.e., the body. As a consequence, this explanation of how the order of imagination arises out of the order of the intellect also explains why Spinoza's *Ethics* is concerned with the "salvation of our mind" from disappearance, and sheds a significant light on Spinoza's doctrine of the eternity of the mind. By knowing things adequately, namely, the subjective mind fuses with the objective mind and acquires its eternity, in a certain sense. As Spinoza concludes:

> The wise man, insofar as he is considered as such, is hardly troubled in spirit, but being, by a certain eternal necessity, conscious of himself, and of God, and of things, he never ceases to be, but always possesses true peace of mind. (E VP42S; G II, 308)

This theme cannot be developed here, yet it is worth mentioning that the realist, and more "empirical" reading of Spinoza's metaphysics of the finite modes, opens

---

[31] I cannot deal with this question here, but I have shown elsewhere that there is a minimal threshold of complexity of an individual for it to become capable of sensation and consciousness. See the references in footnotes 4 and 36.

up routes of investigation that may be worth mentioning for at least three different areas of inquiry:

1. Concerning the eternity of the mind, this interpretation explains why, at the death of the body, the empirical mind vanishes, and all its passive perceptions and affects with it: there remain only those adequate ideas that have constituted it (and that were contained in God for all eternity as eternal possibilities of existence).
2. Concerning the particular status of humans within nature, the fact that they are endowed (in an exclusive way) with reason enables them to build on their inadequate knowledge in order to reach the adequate one. Any explanation of the adequate kinds of knowledge in humans must then show how they are rooted in imagination, i.e., in the subjective mind.
3. Concerning consciousness and knowledge, the fact that a universal animism is posited at the intellectual level does not mean for that sake that all things have an actually *thinking* mind (in the sense of a sensitive, experiential mind). Spinoza only means that God has an adequate idea of them in his intellect.

Difficulties certainly remain, but surely the two minds can confront them better than one.

References

Allison, Henry E. 1987. *Benedict de Spinoza: An introduction.* New Haven, CT: Yale University Press.

Balz, Albert. 1918. *Idea and essence in the philosophies of Hobbes and Spinoza.* New York: Columbia University Press.

Bayle, Pierre. 1740. *Dictionnaire historique et critique.* 4 vols. Cinquième édition, revue, corrigée, et augmentée, avec la vie de l'auteur, par Mr. Des Maizeaux. Amsterdam: Brunel et al.

Bennett, Jonathan. 1984. *A study of Spinoza's ethics.* Indianapolis, IN: Hackett.

Beyssade, Jean-Marie. 1994. "Sur le mode infini médiat dans l'attribut de la pensée. Du problème (Lettre 64) à une solution (*Éthique* V, 36)," in *Revue philosophique de la France et de l'étranger*, 184(1): 23–26.

Bartuschat, Wolfgang. 1994. "The infinite intellect and human knowledge". In *Spinoza on Knowledge and the Human Mind,* ed. Yirmiyahu Yovel, 187–208. Leiden: E. J. Brill.

Brunner, Constantin. 1968. *Science, spirit, superstition: A new enquiry into human thought.* Abr. and trans., Abraham Suhl; rev. and Ed. Walter Bernard. London: Allen & Unwin.

Brunner, Constantin. 1976. *Materialismus und Idealismus.* Gravenhage: Stichting Internationaal Constantin Brunner Instituut. First ed. 1927.

Curley, Edwin. 1969. *Spinoza's metaphysics: An essay in interpretation.* Cambridge: Harvard University Press.

Curley, Edwin. 1988. *Behind the geometrical method. A reading of Spinoza's Ethics* . Princeton: Princeton University Press.

Della Rocca, Michael. 2008a. Rationalism run amok: Representation and the reality of the emotions in Spinoza. In *Interpreting Spinoza,* ed. Charles Huenemann, 26–52. Cambridge: Cambridge University Press,

Della Rocca, Michael. 2008b. *Spinoza.* New York: Routledge.

Erdmann, Johann Eduard. 1834. *Einleitung in eine wissenschaftliche Darstellung der Geschichte der neuern Philosophie.* Riga: Frantzen.

Gueroult, Martial. 1968. *Spinoza.* Vol. I: *Dieu.* Paris: Aubier.

Hegel, Georg Wilhelm Friedrich. 1974. *Lectures on the history of philosophy.* Vol. 3. London: Routledge and Kegan Paul.

Jaquet, Chantal. 2005. "L'amour du corps dans le *Court traité*". In *Spinoza, philosophe de l'amour,* eds. Chantal Jaquet, Pascal Sévérac, and Ariel Suhamy, 45–51. Saint-Étienne: Publications de l'Université de Saint-Étienne.

Joachim, Harol H. 1901. *A study of the ethics of spinoza.* Oxford: Oxford University Press.

Lloyd, Genevieve. 1994. *Part of nature: Self-knowledge in Spinoza's ethics.* Ithaca: Cornell University Press.

Malebranche, Nicolas. 1997. *Dialogues on metaphysics and on religion,* Ed. Nicholas Jolley, trans. David Scott. Cambridge: Cambridge University Press.

Malinowski-Charles, Syliane. 2004. *Affects et conscience chez Spinoza: L'automatisme dans le progrès éthique.* Hildesheim: Olms.

Malinowski-Charles, Syliane. 2009. Le statut psychophysique du plaisir et de la douleur chez Spinoza. Forthcoming in *Historia Philosophica* 5.

Nyden–Bullock, Tammy. 2007. *Spinoza's Radical Cartesian Mind.* London/New York: Continuum.

Pollock, Frederick (Sir). 1899. *Spinoza: His life and philosophy.* London: Duckworth and Co.

Scruton, Roger. 1986. *Spinoza.* Oxford: Oxford University Press.

Spinoza, Benedict. 1925. *Spinoza Opera,* Ed. Carl Gebhardt. 4 vols. Heidelberg: Carl Winters Universitätsbuchhandlung.

Spinoza, Benedict. 1985. *The collected works of Spinoza,* Ed. and trans. Edwin Curley. Vol. 1. Princeton: Princeton University Press.

Spinoza, Benedict. 1994. *A Spinoza reader,* Ed. and trans. Edwin Curley. Princeton: Princeton University Press. [Used for the letters and some excerpts from the *Short Treatise.*]

Wolfson, Harry Austryn. 1934. *The philosophy of Spinoza. Unfolding the latent processes of his reasoning.* 2 vols. New York: Schocken Books.

Part IV
Legacies of Rationalism

Chapter 9
Spinoza's Anti-Humanism: An Outline

Yitzhak Y. Melamed

A triangle, if it could speak, would likewise say that God is eminently triangle, and a circle that God's nature is eminently circular.

- Spinoza, Letter 56

But if cattle and horses or lions had hands, or were able to draw their hands, and do the works that man can do, horses would draw the forms of the gods like horses, and with cattle like cattle, and they would make their bodies such as they each had themselves.

- Xenophanes, Fr. 169[1]

9.1 Introduction

It is because of this that God humanizes himself, that he is willing to allow anthropomorphism, and that he enters into society with us as a prince with his subjects

- Leibniz, Discourse on Metaphysics, §36

Y.Y. Melamed (✉)
Johns Hopkins University, Baltimore, MD, USA
e-mail: ymelame1@jhu.edu

This paper is a first attempt to explore the scope of Spinoza's critique of humanism. Several issues mentioned in this paper are not adequately discussed due to limitations of space. I hope to develop the paper into a larger project in the future. I am indebted to Robert Adams, Hillel Braudie, Michael Della Rocca, Carlos Fraenkel, Zachary Gartenberg, Zeev Harvey, Eve Krakowski, Jean-Luc Marion, Alan Nelson, David Nirenberg, Oded Schechter, Neta Stahl, Lina Steiner, and Peter Thielke for their helpful comments on earlier versions of this paper. Unless otherwise marked, all references to the *Ethics*, the early works of Spinoza, and Letters 1–29 are to Curley's translation (1985; henceforth **C**). In references to the other letters of Spinoza I use Shirley's translation (1995; henceforth **S**). I also rely on Shirley's translation of Spinoza's *Theological-Political Treatise* (2001). Passages in the *Ethics* will be referred to by the following abbreviations: a(-xiom), c(-orollary), p(-roposition), s(-cholium), and app(-endix); "d" stands for either "definition" (when it appears immediately to the right of the part of the book) or "demonstration" (in all other cases). Hence, E1d3 is the third definition of part 1 and E1p16d is the demonstration of proposition 16 of part 1. Occasionally, I will supplement a reference to Gebhardt's Latin edition, by volume, page, and line (hence, II/23/5 is volume II, page 23, line 5). I will use this notation when the reference by proposition number (in the *Ethics*), chapter, or letter is not specific enough.

[1] Kirk, Raven, and Schofield (1983, 169)

C. Fraenkel et al. (eds.), *The Rationalists: Between Tradition and Innovation*,
The New Synthese Historical Library 65, DOI 10.1007/978-90-481-9385-1_9,

I. A common perception of Spinoza casts him as one of the precursors, perhaps even founders, of modern humanism and Enlightenment thought.[2] Given that in the twentieth century, humanism was commonly associated with the ideology of secularism and the politics of liberal democracies, and that Spinoza has been taken as voicing a "message of secularity" (Yovel 1989, 200) and as having provided "the psychology and ethics of a democratic soul" (Smith 2003, 200) and "the decisive impulse to ... modern republicanism which takes it bearings by the dignity of every man" (Strauss 1965, 16),[3] it is easy to understand how this humanistic image developed. Spinoza's deep interest in, and extensive discussion of, human nature may have contributed to the emergence of this image as well. In this paper, I will argue that this common perception of Spinoza is mistaken and that Spinoza was in fact the most radical *anti-humanist* among modern philosophers. Arguably, Spinoza rejects any notion of human dignity. He conceives of God's—and not man's—point of view as the only objective perspective through which one can know things adequately, and it is at least highly questionable whether he allows for any genuine notions of human autonomy or morality.

The notions of "humanism" and "anti-humanism" have been discussed extensively—mainly among continental philosophers[4]—since the end of World

---

[2]For the view of Spinoza's philosophy as anticipating "secularism...,the Enlightenment, and the liberal- democratic state," see Yovel (1989, ix). For the view of Spinoza as a humanist, see Fromm (1964). Any quick search on the web will yield dozens of ideological characterizations of Spinoza as one of the heroes of modern humanism. See, for example, the following declaration: "We, the Sixth International Congress of the IHEU (International Humanist and Ethical Union), representing humanists from all over the world, meeting in Amsterdam on August 5–9, 1974, wish to pay special tribute to Benedict de Spinoza...Spinoza is one of the greatest forerunners of humanist philosophy in modern time. A defender of intellectual and religious liberty and the free mind, he attempted to establish ethics on rational foundations independent of religious dogma. Standing as a bridge between the Middle Ages and Modern science, Spinoza was committed to the use of reason as a source of human freedom" (International Humanist and Ethical Union). Similarly, in a declaration signed by an impressive group of philosophers and intellectuals (among them, W.V. Quine, Arthur Danto, Ernst Nagel, George Hourani, Sidney Hook, Walter Kaufman, and A.J. Ayer) Spinoza is included in a list of "distinguished secularists and humanists who have demonstrated moral principles in their personal lives and works" (Council for Secular Humanism).

[3]On Spinoza as a champion of human dignity, see also Smith (1997, xvi): "Spinoza did not use the term 'liberal' to describe his system of politics .... But if to be a liberal means to have a lively sense of the autonomy and dignity of the individual, ... then Spinoza can be described as a liberal."

[4]Nietzsche, Heidegger, Althusser, and Foucault are probably the most prominent philosophers associated with anti-humanism, though at least in the case of Heidegger, the appropriateness of this association is, to my mind, questionable. In his "Letter on Humanism" (1947, 204), Heidegger criticizes the traditional understanding of the essence of man as '*animal rationale.*' According to Heidegger, this definition fails to recognize man's unique relationship with language and Being. "Only man is admitted to the destiny of ek-sistence. Therefore ek-sistence can also never be thought of as a specific kind of creature among others" In this sense, Heidegger is an arch-humanist. Furthermore, Spinoza seems to be much more radical than Nietzsche in his critique of humanism. The two share a significantly similar conception of good and evil and are both strict naturalists. Yet, Nietzsche never goes beyond the relativity of human perspectives. For Spinoza, there is an objective perspective, but it is God's. Max Black is one of the very few analytic philosophers who have developed a serious interest in the issue of humanism. See his (1983).

War II. Because these notions carry a variety of historical, ideological, and philosophical meanings, it is important to provide at the outset at least a rudimentary clarification of *my* use of these two terms. By "humanism" I mean a view which (1) assigns a *unique value* to human beings among other things in nature, (2) stresses the primacy of the human *perspective* in understanding the nature of things, and (3) attempts to point out an essential property of humanity which *justifies* its elevated and unique status.[5] This definition of *philosophical* humanism has only little in common with the *historical* notion of Renaissance humanism,[6] and seems to match quite well the common understanding of philosophical humanism suggested by current philosophical dictionaries and encyclopedias.[7]

This notion of humanism should be understood in contrast to two competing positions. On the one hand, in contrast to the *theocentric* position that considers humanity to be radically dependent upon God, humanism affirms at least some degree of human independence. On the other hand, in contrast to the *naturalist* position which endorses the scientific examination of human beings just like any other objects in nature, humanists affirm the existence of a metaphysical and moral *gulf* between humanity and nature. This gulf assigns a special value to humanity and does not allow us to treat human beings like any other things in nature. For many humanists, the nature/humanity gulf does not allow the application of the methods of natural sciences to the disciplines of the humanities.[8]

Humanism does not begin with modernity. In order to see how far back we can trace this position, we may recall Protagoras' saying: "Man is the measure of all things, of things that are, that they are, and of things that are not, that they are not."[9] In modern philosophy, the humanistic position had regained dominant status

---

[5]This definition of humanism is intended to be wider than the ideology of secular humanism (pointed out at the beginning of this paper) in order to include religious humanist philosophies like that of Leibniz. Obviously, by claiming that Spinoza was an anti-humanist, I take him to be an enemy of both secular and religious humanism. On the other hand, *speciesism* (the view which suggests that we should favor human beings *only* by virtue of their belonging to our species) would not count as humanism for our purposes. It is not hard to detect the Kantian undertones of my definition of 'humanism,' though a very similar view is expressed by Max Black: "[I]n calling human beings *persons*, we are rightfully ascribing to them important properties that cannot, even in principle, apply to other animals or to inanimate material beings" (Black 1983, 99). According to Black, self-consciousness is such a distinctive characteristic of human beings (Black 1983, 104).

[6]The Renaissance humanism of Lorenzo Valla, Erasmus, and Reuchlin has much more to do with the revival of the *studia humanitatis* than with the glorification of man (though admittedly, these were not completely separate). Giovanni Pico Della Mirandola is almost the only figure of Renaissance humanism who is clearly a champion of the philosophical humanism I define above.

[7]See, for example, Audi (1995, 396–7).

[8]Some famous proponents of the latter view are Wilhelm Dilthey and the Neo-Kantian philosophers, Wilhelm Windelband and Heinrich Rickert. Here again, Max Black provides a crystal-clear statement of this position: "I believe that there are features of human personality that are outside the purview of any of the natural or social sciences, and that there is something therefore conceptually—or, if you like, ontologically—special about human beings" (Audi 1995, 99).

[9]Plato, *Theaetetus* 152a. The "Ode to Man" in Sophocles' *Antigone* (lines 332–375) is another important statement of humanism in ancient Greek culture.

since the Renaissance, and variants of this position were vigorously argued for by prominent thinkers such as Pico della Mirandola, Descartes, Leibniz,[10] Kant, Fichte, and finally, Hegel.[11]

In this paper, I will argue that Spinoza was a foe, and not a friend, of this tradition.[12] I suggest that, in contrast to these humanist philosophers, Spinoza considers man as a marginal and limited being in nature, a being whose claims and presumptions far exceed its abilities. "To what length will the folly of the multitude not carry them? [T]hey imagine Nature to be so limited that they believe man to be his chief part."[13] Arguably, Spinoza locates the origin of our most fundamental metaphysical and ethical errors in a human *hubris* which not only tries to secure humanity an exceptional place in nature but also attempts to cast both God and nature in its own human image.[14]

My view of Spinoza as an "anti-humanist" relies on the following four elements of his thought: (1) Spinoza's perception of human beings as rather marginal and limited beings in an infinite universe, (2) Spinoza's critique of anthropomorphism as a baseless arrogance that causes people to believe that the world is arranged to fit their fictions and caprices, (3) Spinoza's radical naturalism about human beings which denies the existence of *any* gulf between humanity and the rest of nature, and, finally, (4) Spinoza's amoralism. Each of these elements has been subject to detailed discussion in the existing literature; however, as far as I know, they have never been taken as fitting together into a comprehensive world-view. The attempt to draw the outline of such a comprehensive world-view is, I believe, the major innovation of the current paper.

I understand "Rationalism"—the theme of the current volume—as a view that commits itself to the explicability of every fact (or, if you wish, to an unreserved acceptance of the Principle of Sufficient Reason).[15] Spinoza's critique of humanism

---

[10]See Leibniz's *Discourse on Metaphysics*, §§34–36 in G.W. Leibniz (1989, 65–68).

[11]For Hegel's observation and critique of the "annihilation of man" in Spinoza, see Hegel (1995, III 282).

[12]Hyppolite (1997, 20) seems to disclose a similar view of Spinoza in noting that "Hegel is still too Spinozistic for us to be able to speak of a pure humanism." Althusser (1976, 136) too detects some anti-humanist elements in Spinoza by pointing out Spinoza's "radical criticism of the central category of imaginary illusion, the *Subject.*" I discuss the so-called "elimination of the self" in § IV below. Althusser saw, however, only the tip of the iceberg, and the picture I attempt to draw in this paper is far wider and more substantial. In general, Althusser's reading of Spinoza, while occasionally insightful, is quite crude and ideologically biased. See, for example, his ascription to Spinoza of a causality "which would account for the action of the Whole on its parts, and of the parts on the Whole—an unbounded Whole, which is only the active relation between its parts" (Althusser 1989, 141). If I am not mistaken, "the Whole" in question is Spinoza's substance, but the latter is neither acted on by its parts, nor is the activity of substance "the active relation between its parts." For Spinoza, substance is strictly indivisible (E1p13).

[13]*Theological-Political Treatise*, Ch. 6 (III/82).

[14]"The human intellect is deceived simply by its own nature, and feigns everything from the analogy of its own nature, not from the analogy of the universe" (Ep. 2 (IV/8/33)).

[15]For a similar understanding of rationalism (and an interpretation of Spinoza's philosophy that takes his rationalism to be the core of the system), see Michael Della Rocca's recent book (2008).

is at least partly motivated by his strict rationalism. The demand for thorough explicability does not allow for any view which assumes the inner value of humanity as a brute fact. But rationalism does more than that. Rationalism rejects the existence of any "islands" within nature which are governed by "special" laws. As we will soon see, this rejection of human "dominion within dominion" leads Spinoza to debunk any attempt to identify some unique human quality that endows humanity with dignity.

In the first part of this paper, I will attempt to sketch some of the outlines of Spinoza's philosophy, point out the limited place of humanity within this universe, and present Spinoza's views on some crucial issues such as human freedom, human self-knowledge, and the nature of the human mind. In the second part, I will discuss several aspects of Spinoza's critique of anthropomorphism. In the third part, I will discuss Spinoza's naturalistic account of human beings and his amoralism.

It is important to note that by describing Spinoza's philosophy as anti-humanist I do *not* mean to suggest that it either despises[16] or is indifferent to human affairs. As I have already mentioned, Spinoza was deeply interested in the question of human nature. Indeed, the greatest bulk of Spinoza's texts deal with the nature of human beings, their associations, and their illusions. However, his interest in human beings results not from any admiration of man or from a belief in the exceptional place of human beings in the universe, but rather from the very simple fact that Spinoza himself was a human being, and that he considered an illusion-free understanding of humanity to be necessary for understanding his own life and "its highest blessedness."

9.2 The Place of Humanity in Spinoza's World

> [For Spinoza], the human mind is but a light-ray of infinite thought; the human body is but a particle of infinite extension.
>
> - Heine, Religion and Philosophy in Germany.[17]

II. *Man's Marginality in Spinoza's Universe*. At the beginning of the *Ethics*, Spinoza defines God as a "substance consisting of an infinity of attributes [*substantiam constantem infinitis attributis*]" (E1d6). Human beings, however, are constituted by pairs of modes of two of God's infinitely many attributes, Thought and Extension. It is only these two attributes which we can know.

> We neither feel nor perceive any singular things except bodies and modes of thinking (E2a5).

For Spinoza, a human being is simply a pair of two modes of God: a body (a mode of extension) and an idea of that body (i.e., a mind): "[M]an consists of a mind and

[16] See E4p35s (II/234) and E4app13 (269–70).

[17] Heine (1985, 175)

a body" (E2p13c). Yet, since Spinoza's doctrine of parallelism holds that the order of modes is the same in all attributes, and that modes parallel to each other are identical, one may wonder why we are unable to know the modes parallel to our body in all the other attributes. If I know my body, and a certain mode of the third attribute is *identical* with my body, how can I *not* know it? When challenged with this question by one of his correspondents, Spinoza replied that the mode of the third attribute, which is identical with my body, has its own idea, or mind. However, Spinoza adds, the mind of my body and the mind of the mode of the third attribute (which is identical with my body), "cannot constitute one and the same mind of a particular thing. For each of these ideas has no connection [*nullam connexioinem*] with the others" (Letter 66).

Since the mind of my body and the mind of the mode of the third attribute, which is identical with the body, have "no connection with each other," it makes sense that the two minds cannot know each other. Yet it is still not clear *why* these two minds have "no connection with each other." This enigmatic doctrine has caused much controversy and wonder among Spinoza scholars. I believe it is a serious doctrine that is consistent with the rest of his system, though space does not permit full discussion of this doctrine in the present paper.[18] Nevertheless, I think we can already recognize here a huge gap between the infinity of God/Nature, and the limitedness of human knowledge, which captures merely two attributes.[19] It is not only that the human mind can never grasp the vast majority of the infinite attributes, but also, in a sense, that the human being is limited in its ability to know itself. My body has infinite parallel modes in these infinitely many unknown attributes. These modes are identical with my body, and in some odd sense, they are me (conceived under other attributes).[20] Yet, I have no idea what they are and what they do within the infinitely many attributes unknown to the human mind (Only God knows what they do there!).

If the Copernican Revolution threw man from the center of the physical universe, Spinoza's metaphysics multiplied this fall infinitely. It is not just that humans are no longer at the center of the (extended) world, but that even the extended world itself turns out to be just one of an infinite number of aspects of nature, the rest of which are eternally barred from human grasp.

III. *Self-Knowledge*. Spinoza does allow for the human mind to know itself. Furthermore, all of my knowledge of external bodies is mediated through my knowledge of my own body.[21] Yet, this is a rather poor and unreliable form of

---

[18]For a detailed explanation of this issue, see §5.1 of my forthcoming book.

[19]In fact, the human mind can know only *one* attribute (extension) and a *tiny* aspect of the attribute of thought (i.e., ideas which represent bodies).

[20]Cf. Joel Friedman (1983, 105): "It follows, again contra Descartes, that I am much more than a thinking thing."

[21]See E2p113 and E2p16. For a very helpful discussion of these passages see, Michael Della Rocca (1996, 24–29, 47–48, and 64–66). The role Spinoza assigns to self-knowledge in the attainment of blessedness (E5p15) is primarily a result of the fact that almost all of our knowledge is mediated through our knowledge of ourselves.

self-knowledge:[22] "The Mind does not know itself, except insofar as it perceives the ideas of the affections of the Body" (E2p23). For Spinoza, human self-knowledge consists merely of our ability to have first and second order ideas (ideas of our bodies and of our minds, respectively). If I have an idea of a certain event that occurred in my finger, I would also have an idea of that idea. That is all that self-knowledge amounts to for Spinoza.[23] These second order ideas have no privileged characteristics, such as clarity or certainty. In fact, Spinoza claims that "the idea of the idea of any affection of the human Body does not involve adequate knowledge of the human Mind" (E2p29). The inadequacy of human self-knowledge is even more striking given Spinoza's view that human beings have an adequate knowledge of God's essence.[24] Thus, it would seem that, for Spinoza, my knowledge of God's essence is *more adequate* than my knowledge of myself.

While for other modern philosophers, such as Descartes, self-knowledge was both the most certain and the most fundamental knowledge,[25] it has no such privileged status for Spinoza. Furthermore, as I will later argue, Spinoza does not seem to limit self-knowledge to human beings.

IV. *The Non-Substantiality of the Human Mind.* Another common characteristic of the human mind in modern philosophy is its independent existence as a *substance*, a view which was held by Descartes, Leibniz, Locke, and Berkeley, to name a few. Here, again, we find Spinoza in striking opposition to the dominant view. For Spinoza, the human mind is neither a substance nor even a genuinely *simple* being, but rather a mere functionally unified collection of ideas (i.e., modes of Thought):

> The being of substance does not pertain to the essence of man, or substance does not constitute the form of man [*substantia formam hominis non constituit*] (E2p10).

> The idea that constitutes the formal being of the human Mind is not simple, but composed of a great many ideas [*Idea, quae esse formale humanae mentis constituit, non est simplex, sed ex plurimis ideis composita*] (E2p15).

This denial of the substantiality of the human mind has two crucial implications. First, it denies the existence of a thinking subject that is anything more than

---

[22]Several scholars have ascribed to Spinoza a much stronger view of self-knowledge. See for example, Alan Donagan (1998, 117): "[N]o reflective human being ... can fail to perceive that the idea of himself as thinking cannot be false." As I will argue shortly, I do not think Spinoza shared this Cartesian view.

[23]Though Lia Levy has recently presented a very interesting attempt to reconstruct a thicker account of self-consciousness in Spinoza based on Spinoza's discussion of the affects.

[24]E2p47: "The human mind has an adequate knowledge of God's eternal and infinite essence." Cf. E247s: "God's infinite essence and his eternity are known to all."

[25]Recall Descartes' memorable conclusion of the Second Meditation: "I know plainly that I can achieve an easier and more evident perception of my own mind than of anything else" (*The Philosophical Writings of Descartes*, trans. John Cottingham, Robert Stoothoff, and Dugald Murdoch (Cambridge: Cambridge University Press, 1984), vol. 2, pp. 22–3 (AT VII 34)).

a bundle of ideas.[26] Second, it rejects the Cartesian ascription of unique ontological independence to human minds.[27]

Similarly, Spinoza claims that the human body is not a substance but is rather a collection of modes of extension.

V. *The Denial of Free Will.* Descartes's affirmation of the freedom of the will is the target of one of Spinoza's sharpest criticisms.[28] For Spinoza, the notion of free will [*voluntas*] is nothing but a human illusion which results from the fact that human beings "are conscious of their actions and ignorant of the causes by which [their actions] are determined" (E2p34;s).[29] In fact, Spinoza denies that *any* being—even God (E1p32c1)—has free will, since everything that happens, happens necessarily.

Yet, in spite of his unequivocal denial of free will, Spinoza does not repudiate all notions of freedom. At the beginning of the *Ethics*, Spinoza defines a free thing as follows:

> That thing is called free [*libera*] which exists from the necessity of its nature alone, and is determined by itself alone. But a thing is called necessary, or rather compelled [*coacta*], which is determined by another to exist and to produce an effect in a certain and determinate way (E1d7).

This definition allows Spinoza to claim that,

> Although God exists necessarily, he nevertheless exists freely because he exists solely from the necessity of his own nature (Letter 58),

and that

> God alone is a free cause [*causam liberam*]. For God alone exists only from the necessity of his nature, and acts from the necessity of his nature (E1p17c2).

Since, like any other finite beings, human beings do not exist "from the necessity of their nature alone," and since their actions are always determined by external

[26] A crucial implication of the non-simplicity of the human mind is that it seems to undermine one of the most common arguments for mind eternity. Indeed, Van Blijenbergh, one of Spinoza's correspondents and a Christian Cartesian, attacks Spinoza on precisely this point. If in death, claims Van Blijenbergh, "as the human body, when it disintegrates, is resolved again into the thousands of bodies of which it was composed, so also our mind.... And as the scattered bodies [which composed] our human body no longer remain bound to one another, but other bodies separate them, so also it seems to follow that, when our mind is disintegrates, those countless thoughts of which it was composed are no longer combined, but separated" (Spinoza 1985, Ep. 24, 391).

[27] Cf. Van Blijenbergh's complaint in Ep. 20 that Spinoza "makes man dependent on God in the way the elements, stones, and plants are" (IV/103/15). For the substantiality of the human mind in Descartes, see the Second Meditation. For Descartes's definition of substance, see his *Principles of Philosophy*, I 51.

[28] See, e.g., E2p33s2, Letter 21, Letter 58.

[29] Cf. Letter 58 (S 284): "[T]hat human freedom which all men boast of possessing ... consists solely in this, that men are conscious of their desire and unaware of the cause by which they are determined. In the same way a baby thinks that it freely desires milk, an angry child revenge, and a coward flight." See Michael Della Rocca, "The Power of an Idea: Spinoza's Critique of Pure Will," *Nous* 37 (2003), 200–231.

causes as well, it is clear that they cannot be free even according to Spinoza's own definition of freedom.[30] However, in various places in his writings Spinoza attempts to mitigate this conclusion by relaxing his definition of freedom so that it would allow for a variety of *degrees* of freedom.

In his correspondence with Van Blijenburgh, Spinoza seems to hold that *the more* our actions follow from our nature, *the more* we are free:

> If God's nature is known to us, then affirming that God exists follows necessarily from our nature .... [W]e are never *more free* than when we affirm a thing in such a way (Letter 21| IV/130/6; italics mine).

Since Spinoza holds that the actual essence, or nature, of every finite being is the striving to persevere in its being (E3p8), it would seem that the more a person strives to persevere in her being, the more she is free.

The result of this relaxation of the definition of freedom is a rather humble view of human freedom. Furthermore, freedom does *not* seem to distinguish man from the rest of nature. The *conatus*, or the striving to persevere in one's being, belongs to the essence of any finite being,[31] and to that extent, any cockroach which follows its essence and strives to persevere in its being is—to some extent—free.[32]

9.3 The Battle Against Anthropomorphism.

VI. *The Finite and the Infinite.* In E2p10s2, Spinoza rebukes those philosophers who

> did not observe the order of Philosophizing. For they believed that the divine nature—which they should have contemplated before all else (*because it is prior both in knowledge and in nature*)—is last in the order of knowledge, and the things that are called objects of the senses are prior to all. That is why, when they contemplated natural things, they thought of nothing less that they did of the divine nature; and when afterwards they directed their minds to contemplating the divine nature, they could think of nothing less than their first fictions [*figmentis*]." (Emphasis mine)

That philosophy must begin with the infinite and then conceive the finite through the infinite (and not the other way around) is one of Spinoza's most important

---

[30] For an insightful discussion of the "free man" as an impossible model, see Dan Garber (2004, 183–207).

[31] E3p6: *Unaquaeque res, quantum in se est, in suo esse perseverare conatur.* In his aforementioned article (1983, 105–6), Max Black suggests that having a need for the individual's survival (rather than survival of the species) is another unique characteristic of human beings; he then quotes Spinoza's *conatus* doctrine in support of this claim. It is hard to understand what made Black think that the *conatus* is particularly human, while Spinoza explicitly states "Each thing [*Unaquaeque res*] ...." Indeed, in Letter 58, Spinoza openly discusses the *conatus* of the stone which "as far as in it lies" strives "to continue in motion" (S 284).

[32] Spinoza's rejection of free will seems to motivate his views on the punishment of criminals. Since Spinoza does not consider harmful actions by human beings to be anchored in free and morally responsible agents, he suggests that the punishment of criminals should be justified by the very same considerations which make people exterminate poisonous snakes (CM II, viii| I/265/23).

methodological principles and arguably one of his most significant innovations.[33] This principle is expressed most clearly in the definitions of Substance and Mode (E1d3&5) that make the modes dependent on substance both for their existence and for their conceivability. It also functions as the conceptual foundation of Spinoza's crusade against anthropomorphism.

It is not easy to summarize all the errors, illusions, and misconceptions that Spinoza finds to result from anthropomorphic thinking. In the appendix to the first part of the *Ethics*, Spinoza lists *good*, *evil*, *order*, *confusion*, *warm*, *cold*, *beauty*, *ugliness* as notions that result from people's belief that "everything that happens, happens on their account," and that everything is valued by its usefulness for human beings (II/81/27). This list is far from exhaustive.

Spinoza presses the charge of anthropomorphism against both philosophers and theologians, both Jews (whom he describes as the most "accustomed to grant all things human attributes"[34]) and Christians (whom he criticizes for believing that "God took upon himself human nature"[35]). Here, I will concentrate on four issues that seem to me the most crucial for Spinoza's battle against anthropomorphism: the critique of Scripture, teleology, the problem of evil, and the critique of morality.

VII. The Critique of Scripture. The claim that Scripture describes God in vulgar, anthropomorphic terms prevails both in the *Theological Political Treatise* and in Spinoza's correspondence.[36] In fact, however, this is the *least* innovative aspect of Spinoza's discussion of anthropomorphism. Indeed, Spinoza explicitly attests: "I have never seen a Theologian so dense [*crassum*] that he did not perceive that Sacred Scripture very often speaks of God in a human way" (Letter 21 | IV/132/22). The claim that Scripture frequently describes God in human terms in order to be accessible to the masses was not only a major theme in medieval Jewish philosophy (especially for Maimonides and his followers[37]), but it had also been a central hermeneutic principle in the early Talmudic literature, which relentlessly warns the reader that "Scripture speaks in the language of common human beings."[38] In this respect, Spinoza's biblical criticism contributes little that is new.

VIII. *The Critique of Teleology*. At the beginning of the Appendix to Part One of the *Ethics*, Spinoza states:

> All the prejudices I here undertake to expose depend on this one: that men commonly suppose that all natural things act, as men do, on account of an end; indeed, they maintain as

---

[33]Cf. KV I, 22 (I/101/3–7) and KV II xxiv (I/107/1). For Spinoza's critique of those who claim to know God only through created things, see TTP Ch. 2 (III/30).

[34]Spinoza (1962, 29).

[35]Letter 73.

[36]See, for example, the first two chapters of the TTP and Spinoza's correspondence with Van Blijenbergh (Letters 18–24).

[37]See Moses Maimonides, *Guide of the Perplexed*, Part I, Chapter 26. For an excellent and comprehensive account of Maimonides' influence on Spinoza (including the critique of anthropomorphism), see Warren Zeev Harvey (1981, 151–72 [esp. 164]).

[38]See, for example, *Babylonian Talmud*, *Tracatate Hulin*, 4b; cf. *Tractate Nedarim*, 3a, and *Tractate Avoda Zara*, 27a.

> certain that God directs all things to some certain end, for they say that God has made all things for man, and man that he might worship God (II/78/1–6).

Being aware of their appetites—and ignorant of the causes of these appetites—people believe that their actions are freely determined by the purposes they set for themselves (E1app (II/78/18–22) and E4Pref (II/206–7)).[39] They also believe that other intelligent beings act in a similar way. When they find various things in nature that are useful for them and know that these things were not created by human beings, they feign the existence of gods and assume that the gods created those things for the sake of man. Being ignorant of the nature of these gods, people assume that they are just like them, though much more powerful. Thus, they try to satisfy the gods with gifts, offerings, and prayers (E1App). This, *in nuce*, is Spinoza's analysis of the genealogy of religion.

For Spinoza, this kind of thinking is a mixture of illusion, ignorance, and *hubris*. *Deus sive Natura* acts and exists for the sake of no end (E4Pref| II/207/1). Of course, human beings can use nonhuman individuals for their own advantage, but these nonhuman individuals can just as well use humans for their sake. For a human being to eat a cow is no more *natural* than for a lion to devour a human being, or even for one human being to eat another. None of these individuals was created for the sake of another.

For Spinoza, the most crucial error of teleological thinking seems to be the inversion of the infinite-finite relation. When people believe that God acts for the sake of man and in order to be praised by man, they make the infinite depend on the finite; furthermore, to the extent that God, allegedly, acts in order to attain something he lacks, teleological thinking makes the infinite God—imperfect:

> This doctrine concerning the end turns nature completely upside down. For what is really a cause, it considers as an effect, and conversely. What is by nature prior, it makes posterior. And finally, what is supreme and most perfect, it makes imperfect (E1App| II/80/10–14).

IX. *The Problem of Evil*. Spinoza has a simple and clear-cut solution to the problem of evil that at first may appear similar to traditional theodicy, such as Leibniz's: the idea of evil is merely a result of the limitedness of human thinking. But when we ask Spinoza to elaborate, we receive a response radically different from any traditional theodicy. According to Spinoza, when human beings say that a certain particular—whether it is an act, a person, or an event—is evil, what they actually do is to compare the particular in question with a certain perfection it *could* have and judge that it *could have been* better. To put the same idea in Spinoza's own words, evil "is only a privation of a more perfect state" (Letter 19| IV/91/3). In order to judge that the particular in question could have been more perfect, we compare the particular to a universal under which it falls. Realizing that the universal is much better ("more perfect"), we judge the particular to be evil. However, Spinoza

[39] Notice that for Spinoza the rejection of necessity and the belief in free will are preconditions for the emergence of teleological thinking (E1app|II/78/21). The issue of teleology in Spinoza has recently been a subject of intensive debate. Unfortunately, I cannot weigh in here on this important question.

argues, the use of universals is a mark of the limitedness of human cognitive faculties. Universals belong to a certain cognitive compensation mechanism that helps us find our way in the world in spite of the strict limitations of our perception and memory, which cannot conceive a great number of particulars with all their characteristics. Thus, for example, when we see seven flying monsters, we form the concept of a "flying monster" in order to avoid the difficult task of conceiving these seven particulars with all their detailed characteristics.[40] When *God* thinks of particulars he does not conceive them through these abstract universals, but rather knows them directly in their particularity (Letter 19| IV/92/1; Cf. E4Pref| II/207/19).[41] Knowing that the particular in question could not act otherwise, God does not judge it to be lacking anything that would naturally belong to it. Thus, Spinoza argues that privation and evil "can be said only in relation to our intellect, not in relation to God's" (Letter 19| IV/92/20). From God's perspective, says Spinoza, appetition for the good belongs to the nature of a wicked person no more than it belongs to the nature of a stone; hence, neither the stone nor the wicked person is *deprived* of goodness (Letter 21| IV/129/1). In other words, for Spinoza, "evil"—as privation of the perfection of goodness—cannot be attributed to Hitler any more than to a rock.

Had Spinoza witnessed the great earthquake in Lisbon, his response would have been straightforward and simple: there was *nothing* evil in that event. The land of the city was just exhibiting its particular nature, and it was *as perfect as it could be*. To say that the mass death that resulted from that event was evil is to make the erroneous comparison between a particular ("the land of Lisbon," or a certain person who perished in that event) and a universal ("land," or "humanity," respectively), and falsely conclude that the particular lacks a perfection which naturally belongs to it.

X. Spinoza's Amoralism. Given the title of Spinoza's main work[42] and the fact that a considerable part of the book deals with the improvement of human conduct, one may be surprised to find Spinoza described as an "amoralist". Nevertheless, this title is recurrently ascribed to Spinoza, and, I believe, rightly so.[43] For Spinoza's "moral theory" is essentially nothing but a *theory of prudence*. It begins with a clear egoistic foundation and proceeds to show that a prudent egoist would in many respects behave in a way that would be *judged* righteous by common morality, and that he would adopt characteristics that fit the common understanding of

---

[40]For Spinoza's account of universals, see E2p40s1. Cf. TdIE §99 (II/36/18) and CM I, I (I/235/22–5).

[41]"We, on the contrary, attribute knowledge of singular things to God, and deny him a knowledge of universals, except insofar as he understands human minds (*Cogitata Metaphysica*, II, vii [I/263/9]).

[42]Since Spinoza's *summum bonum* is nothing but the knowledge of God (E4p28), it seems that his ethical discussion begins with the very first definition of Part 1 of the *Ethics*; hence the aptness of the title of the work.

[43]Spinoza himself clearly expected the charge of amoralism to be brought against him. See E4p18s (II/223/21–4): "I have done this to win, if possible, the attention of those who believe that this principle—that everyone is bound to seek his own advantage—is the foundation, not of virtue and morality, but of immorality."

virtue.[44] Spinoza's is indeed a very peculiar kind of egoism. One may name it "Egoism without Ego," for, as we have already seen, Spinoza rejects the robust unity of the self.[45] Yet, notwithstanding the fuzziness and weakness of the Spinozistic self, Spinoza encourages each entity to concentrate on the promotion of its own true good.

As I have already mentioned, in the Appendix to Part One of the *Ethics*, Spinoza includes *good* and *evil* in the list of prejudices that result from the belief that man is the end of nature (E1App| II/78/10 and II/81/30). Following this list, Spinoza provides a short elucidation of the common understanding of good and evil: "What conduces to health and the worship of God, they have called *good*; but what is contrary to these, *evil*" (II/81/35). Spinoza's own definition of good and evil—though it excludes the relevancy of the worship of God—does not radically differ from the above common understanding of these two notions:

> By Good I shall understand what we certainly know to be useful to us [*Per bonum id intelligam, quod certo scimus nobis esse utile].*
>
> By Evil, however, I shall understand what we certainly know prevents us from beings masters of some good [*Per malum autem id, quod certo scimus impedire, quominus boni alicuius simus compotes*] (E4d1 & 2).[46]

Along the same line that takes *good* and *evil* (as commonly understood) to be human prejudices, Spinoza includes *just* in the list of anthropomorphic predicates that we erroneously ascribe to God.[47]

The most fundamental doctrine of Spinoza's "moral theory" appears already in the third part of the *Ethics*. This is the famous doctrine of the *conatus*: "The striving [*conatus*] by which each thing strives to preserve in its being is nothing but the actual essence of the thing" (E3p7). In the fourth part of the *Ethics*, Spinoza relies on the doctrine of the *conatus* in order to claim that since the *conatus* is the essence of man, human *virtue* is nothing but human *power* (E4d8 and E4p20d). Thus, the more a person strives to persevere in his being—and does so *prudently*—the more he is endowed with virtue. The more an act strengthens a person's power to preserve herself, the better this act is. In the *Theological Political Treatise*, Spinoza embarks on a very similar line in claiming that an individual's right—no matter whether this individual is a human being, an animal, or a state—extends as far as its power does.[48]

---

[44] For a helpful discussion of Spinoza's egoism, see Della Rocca (2004).

[45] For a detailed discussion of the weakness of individuation in Spinoza, see my article, "Acosmism or Weak Individuals? Hegel, Spinoza, and the Reality of the Finite" (2009).

[46] See E2p29s for equally self-centered definitions of 'praise' and 'blame': "*The Joy with which we imagine the action of another by which he has striven to please us* I call *Praise.* On the other hand, *the Sadness with which we are averse to his action* I call *Blame.*"

[47] *Theological Political Treatise*, Ch. 4 (III/65). Cf. E4p37s2 (II/239). For a compelling argument regarding Maimonides' influence on Spinoza's conception of good and evil, see Harvey (1981, 158–60).

[48] See *Theological Political Treatise*, Preface (III/11), Chapters 16 (III/189) and 20.

It is important to note that Spinoza does not limit the identification of right and power only to the state of nature. Even in the political state, the supreme right of the sovereign is nothing but a reflection of its supreme power.

> With regard to political theory, the difference between Hobbes and myself, which is the subject of your inquiry, consists in this, that I always preserve the natural right in its entirety, and I hold that the sovereign power in a State *has right over a subject only in proportion to the excess of its power over that of a subject.*[49]

When Spinoza advises the sovereign to promote the wellbeing of all his subjects,[50] he is merely counseling *prudence*. A state that promotes the wellbeing of all subjects, Spinoza thinks, is more stable and less likely to go through internal turmoil. Therefore, a prudent sovereign will seek to promote the general good—or at least will make the impression that he does so—*for his own sake*.[51]

Admittedly, from this very egoistic foundation of his practical philosophy, Spinoza derives many counsels that we would willingly embrace. He suggests that we should treat other human beings—or at least the wisest among them—in a just and honorable way, since in nature there is nothing more *useful* to us than the friendship of other—wise—people (E4p18s). However, think about the following situation. Suppose the commander of an extermination camp was a strict Spinozist. Was he doing anything wrong according to Spinoza?[52] It was within his *power* to kill 10,000 people a day and therefore this act would seem perfectly within his *right*. Of course, one may say that killing 10,000 people is not very prudent: one risks the revenge of the families and friends of those executed,[53] and it is also imprudent because this commander could *use* these thousands of people (or the wisest among them) as friends.[54] But is this really what we mean when we say that it was wrong to

---

[49]Letter 50 (IV/238–9); italics mine. Cf. *Theological Political Treatise*, Chapter 16, p. 177, and Chapter 20, p. 223.

[50]See *Theological Political Treatise*, Ch. 16 (III/194) "In their own interest and to retain their rule, it especially behooves [the governments] to look to the public good."

[51]See Curley's reading of Spinoza as the "most Machiavellian of the great modern political philosophers" (1996, 315). Cf. Spinoza's own sympathetic evaluation of Machiavelli (*Poiltical Treatise*, Ch. 5| III/296–7)

[52]In E4p45d Spinoza claims that it is evil to destroy a man we hate. This is so, however, primarily because hate (which is a kind of sadness for Spinoza), is in itself evil.

[53]At the beginning of the 20th Chapter of the TTP (III/240), Spinoza argues that it is not within the right of a tyrant to liquidate his citizens for the most trivial reasons, since such a behavior will put the government in great risk. Thus, Spinoza continues, such behavior is not within the absolute power of the sovereign (and therefore not within his right). While this political rule of thumb may work in many, perhaps most, cases, Spinoza fails to provide any strict proof ruling out the possibility that on certain uncommon occasions it may well be within the sovereign's interest to eliminate some or all of his citizens in order to secure his well being. It is not difficult to construct various scenarios of this sort.

[54]And finally, this mass killing might be imprudent, and hence evil, because society imposes certain emotions on its members, so that they will be mentally tormented by guilty feelings were they to conduct illegitimate killings. Yet, these considerations remain well within the domain of the egoistically imprudent.

be a commander of an extermination camp? Do we merely mean that it was simply imprudent and not conducive to one's own advantage?

9.4 Spinoza's Radical Naturalism

XI. *No Dominion within Dominion.* In the course of modern philosophy, we find a central humanistic stream of thought that attempts to secure humanity a distinguished place, elevated above nature. Spinoza, to my mind, is a foe, not a friend, of this tradition. In the preface to the third part of the *Ethics*, Spinoza famously criticizes those who

> conceived man in nature as a dominion within dominion [*imperium in imperio*]. For they believe that man disturbs, rather than follows [*sequi*], the order of nature, that he has absolute power over his actions, and that he is determined only by himself (II/137/10–15).[55]

Whether it is the capacity to act *freely*, *morally*,[56] *rationally*, or have *self-consciousness*, Spinoza denies that any of these characteristics separate humanity from the rest of nature.[57] Human bodies follow precisely the same laws that govern the body of the snail, and ideas of human bodies (i.e., human minds) are governed by precisely the same laws that govern the mind of the snail.

In the midst of his discussion of the human mind in the second part of the *Ethics*, Spinoza notes:

> The things which we have shown so far are completely general and do not pertain more to man than to other Individuals, all of which, though in different degrees, are nevertheless animate. For of each thing there is necessarily an idea in God, of which God is the cause in the same way as he is of the idea of the human Body. And so, whatever we have said about the idea of the human Body, must also be said of the idea of any thing. (E2p13s| II/96/26–32)

One bold implication of this passage is that snails—and apparently rocks as well—are self-conscious. Since, for Spinoza, self-consciousness is nothing but having a second-order idea of the body, Spinoza would *have to hold* that snails are self-conscious. In the passage above, he states explicitly that all things have minds

[55]Cf. *Cogitata Metaphysica*, II, ix (I/267/10: "[M]an is part of Nature, which must be coherent with the other parts").

[56]In E3p5 to E3p9, Spinoza explicitly claims that "all things"—rocks and hippopotamuses included—have *conatus*. Since his theory of the affects is mostly an explication of the doctrine of the *conatus*, there seems to be no reason why we could not construct similar theories of the affects of rocks and hippopotamuses (alas, the latter would be quite dull). Since the doctrine of the *conatus* and the theory of affects provide the foundations for Spinoza's moral theory, it seems likely that we could even construct a moral theory for hippopotamuses and rocks (provided that Spinoza's "moral theory" for human beings is recognized as a genuine moral theory).

[57]Traditional Aristotelian and scholastic philosophy (with its conception of various kinds of souls) seems to suggest much more continuity between humans and other living beings than the modern philosophies of Descartes and Kant. On this issue, Spinoza seems to be much closer to his medieval predecessors than to most modern philosophers.

(i.e., ideas of their bodies), and since the doctrines of divine omniscience (E2p3) and parallelism (E2p7) commits him to the view that all ideas have their parallel second-order ideas (insofar as ideas too *are* things and must be conceived by one of God's ideas[58]), it seems that all bodies—snails and rocks included—have their own second-order ideas and are thus self-conscious.

A view which states that snails know God's essence and that the snail's mind is eternal may seem even more striking.[59] However, when we look closely at Spinoza's proofs of the doctrines that the human mind has an adequate knowledge of God (E2p45–7), and of the eternity of the human mind (E5p22–23), we see that both proofs rely on very general considerations about the relation of individual minds to God. There seems to be *nothing* in these proofs which is peculiar to the human mind, and apparently nothing that would preclude a construction of similar proofs regarding the snail's—or even the rock's—mind.[60] To view snails and rocks as having adequate knowledge of God as well as eternal minds is indeed quite daring,[61] but it seems to be a clear result of Spinoza's strict naturalism, which denies *any* chasm between human and non-human individuals in nature.

Spinoza does recognize that different particulars have different natures, and though they all strive to persevere in their being, each particular does so in its own way:

> Though each individual lives content with his own nature, by which he is constituted, and is glad of it, nevertheless that life with which each one is content, and that gladness, are nothing but the idea, or soul [*anima*], of that individual. And so the gladness of the one differs in nature from the gladness of the other as much as the essence of the one differs from the essence of the other. (E3p57s)

Spinoza also acknowledges that there is a difference of *degree* between the intelligence of the snail and that of human beings (and between the intelligence of any two individuals, in general). He takes this difference to be the mental parallel to the physical difference between the complexity and capabilities of the snail's body and those of the human body:

> [T]o determine what is the difference between the human mind and the others, and how it surpasses them, it is necessary for us, as we have said, to know the nature of its object, that is, of the human body. I say this in general, that in proportion as a Body is more capable than others of doing many things at once, or being acted on in many ways at once, so its Mind is more capable than others of perceiving many things at once. And in proportion as the actions of a body depend more on itself alone, and as other bodies concur with it less in acting, so its mind is more capable of understanding distinctly. And from these we can know the excellence of one mind over the other. (E2p13s| II/97/7)

---

[58] See E2p20d and its reliance on E2p3 and E2p7.

[59] I discuss this issue in more detail in a yet unpublished paper, "Spinoza on the Fish's Knowledge of God's Essence."

[60] For a similar point, see Wilson (1999, 343).

[61] "It would follow that a tiny, but only a tiny, portion of brutes' minds is 'eternal'. (As much or more as in the case of human babies? Who can tell?)" (Wilson 1999, 350 n.27).

Spinoza relies on the same general principle in order to explain the difference in mental capacities between children and adults (E5p39s| II/305/25). Thus, the difference between man and other animals is of the same kind as (though apparently greater than) the difference between two people or *any* two individuals.[62]

XII. *Spinoza's Speciesism.* Given Spinoza's denial that human beings are qualitatively elevated above the rest of nature, it is interesting to view Spinoza's strong objection to vegetarianism:

> The law against killing animals is based more on empty superstitions and unmanly compassion [*muliebri misericordia*] than sound reason. The rational principle of seeking our own advantage teaches us the necessity of joining with men, but not with the lower animals [*brutis*], or with things whose nature is different from human nature. We have the same right against them as they have against us. Indeed, because the right of each other is defined by his virtue, or [*seu*] power, men have a greater right against the lower animals than they have against men. Not that I deny that the lower animals have sensations. But I do deny that we are therefore not permitted to consider our own advantage, use them at our pleasure, and treat them as is most convenient for us. For they do not agree in nature with us, and their affects are different in nature from human affects. (E4p37s1)

As one can easily see from this passage, Spinoza refuses to invoke any *metaphysical* difference that grants humanity a special value. His argument against vegetarianism is simple and straightforward. Assuming that

1. Human beings are more powerful than other animals (premise),

and that

2. Right is identical with power (premise),

we can conclude that

3. Human beings have more right than other animals (from [1] and [2]).

Having more right than weaker beings "allows" us to use them for whatever purpose we prefer. Yet, since

4. The best way to use another being is to join with it in friendship (premise),

we could have thought that

5. It is best for us to use other animals (or any weak beings) as friends (from [3] and [4]).

Spinoza reminds us, however, that

6. The natures of other animals is significantly different from ours (premise),

and therefore

7. We cannot communicate with animals (from [6]).[63]

But, since

8. Communication is a pre-condition for friendship (premise), it is clear that

---

[62] That Spinoza considered animals to be—to some extent—rational can be seen also from his talk about "the animals which are *called* [*dicuntur*] irrational" (E3p57s| II/187/5; my emphasis).

[63] Although he does not mention it explicitly, Spinoza seems to be relying here on the doctrine of the imitation of affects (E3p27).

9. We cannot use animals as friends (from [6]–[8]).

From (3), (4), and (9) we can finally conclude that

10. Since we cannot use animals as our friends, we can—and may—use them in any other way.

It is not hard to detect here the principles of Spinoza's egoistic "ethics" (seeking one's own advantage) and politics (the identification of right with power). This time, however, they are applied to a group—human beings—rather than to a specific individual. To the extent that this group has any *power* over another group, it has the *right* to use members of the other group "at their pleasure." Fortunately, Spinoza did not believe that there are species of human beings that are superior (even only in degree) to other humans;[64] for otherwise Spinoza would have to be counted among the founders of modern racism. Yet, were I a young child, an autistic person, or even someone who refuses the friendship of B.d.S.,[65] I would think twice before becoming his neighbor.

9.5 Epilogue

XIII. In the current paper, I have not touched upon several important issues that are relevant to Spinoza's critique of humanism. These include Spinoza's reason for supporting democracy, his view of women, his rejection of the separation of state and religion, and the extent to which Spinoza's metaphysics allows him to talk seriously about a unified human nature (i.e., a nature or essence that is shared by all human beings). Also, I have mostly avoided evaluating Spinoza's claims. My aim here was to shatter a popular myth that celebrates Spinoza as a hero of modern humanism and liberalism. The Spinoza I perceive is far darker and more complicated. I do find his criticism of humanistic hubris extremely important and powerful, and there are at least *parts* of this criticism that I would like to accept. But there are also other aspects of his thought (primarily those dealing with morality) that I find quite unsettling. Whether Spinoza's criticism of human hubris can be separated from his view of morality is, I think, a crucial question.

The four aspects of Spinoza view of humanity discussed in this paper—man's marginality, the illusions resulting from anthropomorphic thinking, Spinoza's strict naturalism about human beings, and his amoralism—do not logically necessitate each other; one can consistently adopt each of these positions while rejecting the other three. Yet, they do, I believe, fit with and support each other. Taken together, they constitute a certain comprehensive view, which does not follow from any one of these doctrines alone. This comprehensive view neither ridicules nor eliminates man (as Hegel tends to think),[66] but rather attempts to remind him of his rather humble

---

[64] See *Theological-Political Treatise*, Chapter 3, (III/47).

[65] Indeed, in the TTP (Ch. 16, III/196), Spinoza makes clear that anyone who refuses to be an ally (or citizen) of the state should be considered an enemy against whom any measures may be used.

[66] "[In Spinoza] The world has no true reality, and all this that we know as the world has been cast into the abyss of the one identity. There is therefore no such thing as finite reality, it has no truth

and limited place in nature. In a word, it cuts short any talk of *human dignity*[67] or of the *inner value* of humanity (insofar as this dignity is not shared, to a degree, by snails and rocks as well). A nice way to illustrate this point is by contrasting Spinoza's view of humanity with the opening lines of Kant's *Anthropology from a Pragmatic Point of View* (1798):

> The fact that the human being can have the "I" in his representations raises him infinitely above all other living beings on earth. Because of this he is a *person* [*Person*], and by virtue of the unity of consciousness through all changes that happen to him, one and the same person—i.e., through rank and dignity [*Rang und Würde*] and entirely different being from *things* [*Sachen*], such as irrational animals, with which one can do as one likes. (Kant 2006, 15 [Ak. 7:127])[68]

Hardly a single one of these claims is true, Spinoza would respond. Man is endowed by no special unity of consciousness, no personality or rationality not shared by other beings, and therefore, no preeminence or distinguishing dignity.

References

Althusser, Louis. 1976. *Essays in self-criticism*, trans. Grahame Lock. London: Humanities Press.

Audi, Robert, ed. 1995. *The Cambridge dictionary of philosophy*. Cambridge: Cambridge University Press.

Black, Max. 1983. Humaneness. In *The prevalence of humbug and other essays*, 97–114. Ithaca, NY: Cornell University Press.

Council for Secular Humanism. A Secular Humanist declaration. http://www.secularhumanism.org/index.php?section=main&page=declaration.

Curley, Edwin. 1996. Kissinger, Spinoza, and Genghis Khan. In *The Cambridge companion to Spinoza*, ed. Don Garrett. Cambridge: Cambridge University Press.

Della Rocca, Michael. 1996. *Representation and the mind-body problem in Spinoza*. Oxford: Oxford University Press.

---

whatever; according to Spinoza what is, is God, and God alone. Therefore the allegations of those who accuse Spinoza of atheism are the direct opposite of the truth; with him there is too much God. They say: if God is the identity of mind and nature, then nature or the individual man is God. This is quite correct, but they forget that nature and the individual disappear in this same identity; *and they cannot forgive Spinoza for thus annihilating them*. Those who defame him in such a way as this are therefore not aiming at maintaining God, but at maintaining the finite and the worldly; *they do not fancy their own extinction* ..." (Hegel 1995, vol. 3, 281–2; italics mine). Hegel is fully aware of Spinoza's critique of modern humanism and pays very close attention to most of the issues discussed in this paper. Hegel notes that, in Spinoza's system, there is "an utter blotting out of the principle of subjectivity, individuality, personality, the moment of self-consciousness in Being" (Hegel 1995, vol. 3, 287).

[67] For the contrast between *market price* (relative worth) and *dignity* (inner worth), see Kant's *Groundwork of the Metaphysics of Morals* (Ak. 4:434–5).

[68] Strikingly, Spinoza and Kant, in spite of their opposing views of the value of humanity, arrive at almost the same perspective on the issue of human treatment of animals. Compare Kant's last sentence at the passage above with Spinoza's claim in E4p37s1 that since we cannot use animals as friends we "may use them at our pleasure, and treat them as is most convenient for us." For a recent endorsement of the view that grounds "the distinctive dignity of human beings" in human self-consciousness, See Manfred Frank (1995, 80).

Della Rocca, Michael. 2004. Egoism and the imitation of the affects in Spinoza. In *Spinoza on reason and the "free man,"* eds. Yirmiyahu Yovel and Gideon Segal, 123–47. New York: Little Room Press.
Della Rocca, Michael. 2008. *Spinoza*. New York: Routledge.
Donagan, Alan. 1998. *Spinoza*. Chicago: Chicago University Press.
Frank, Manfred. 1995. Philosophical foundations of early romanticism. In *The modern subject: Conception of the self in classical German philosophy*, eds. Karl Ameriks and Dieter Sturma. Albany, NY: SUNY Press.
Friedman, Joel. 1983. Spinoza's problem of "other minds". *Synthese* 57: 99–126.
Fromm, Erich. 1964. Humanism and psychoanalysis. *Contemporary Psychoanalysis* 1: 69–79.
Garber, Dan. 2004. Dr. Fischelson's dilemma: Spinoza on freedom and sociability. In *Spinoza on reason and the "free man,"* eds. Yirmiyahu Yovel and Gideon Segal. New York: Little Room Press.
Harvey, Warren Zeev. 1981. A portrait of Spinoza as a Maimonidean. *Journal of the History of Philosophy* 19: 151–172.
Hegel, G.W.F. 1995. *Lectures on the history of philosophy*, trans. E.S. Haldane. Lincoln: University of Nebraska Press.
Heidegger, Martin. 1977. Letter on humanism. In *Basic writings*, ed. David Farrell Krell, 213–66. New York: Harper & Row.
Heine, Heinrich. 1985. *The Romantic school and other essays*, eds. J. Hermand and R.C. Holub. New York: Continuum.
Hyppolite, Jean. 1997. *Logic and existence*, trans. Leonard Lawlor and Amit Sen. Albany, NY: State University of New York Press.
International Humanist and Ethical Union. Conference Declaration. http://www.iheu.org/node/771.
Kirk, Geoffrey Stephen, John Earle Raven, and Malcolm Schofield, eds. 1983. *The presocratic philosophers*. 2nd ed. Cambridge: Cambridge University Press; Smith, Steven. 1997. *Spinoza, liberalism and the question of Jewish identity*. New Haven, CT: Yale University Press.
Kant, Immanuel. 2006. *Anthropology from a pragmatic point of view*, trans. Robert B. Louden Cambridge: Cambridge University Press,
Leibniz G.W. 1989. *Philosophical essays*, trans. and ed. Roger Ariew and Daniel Garber. Indianapolis, IN: Hackett.
Melamed, Yitzhak. 2011. *Spinoza's Metaphysics of Substance and Thought*. Oxford: Oxford University Press: forthcoming.
Melamed, Yitzhak. 2010. Acosmism or weak individuals? Hegel, Spinoza, and the reality of the finite. *Journal of the History of Philosophy* 48: 77–92.
Smith, Steven. 2003. *Spinoza's book of life: Freedom and redemption in the Ethics*. New Haven, CT: Yale University Press.
Spinoza, Benedict. 1962. *Hebrew grammar*, trans. Maurice J. Bloom. New York: Philosophical Library.
Spinoza, Benedict. 1985. *The collected works of Spinoza*. Vol. 1. Ed. and trans. Edwin Curley. Princeton: Princeton University Press.
Spinoza, Benedict. 1995. *The letters*, trans. Samuel Shirley. Indianapolis, IN: Hackett.
Spinoza, Benedict. 2001. *Theological-political treatise*. 2nd ed. Trans. Samuel Shirley. Indianapolis, IN: Hackett.
Strauss, Leo. 1965. *Spinoza's critique of religion*. New York: Schocken Books. [Orig. German edition, 1930.]
Wilson, Margaret D. 1999. 'For they do not agree in nature with us': Spinoza and the lower animals." In *New essays on The rationalists*, eds. Rocco G. Gennaro and Charles Huenemann 336–52. Oxford: Oxford University Press.
Yovel, Yirmiyahu. 1989 *Spinoza and other heretics*. Vol. 1. Princeton, NJ: Princeton University Press.

Chapter 10
Spinoza, Leibniz, and the Gods of Philosophy

Steven Nadler

If pressed to name a single issue that was most central to philosophy on the Continent in the second half of the seventeenth century, I would have to say, with little hesitation: theodicy and the problem of evil. A plausible case can be made, I believe, for the claim that the question of the origin of and explanation for the existence of evil, sin and suffering in a world created by an omnipotent, omniscient, wise, and benevolent God—as well as the apparent injustice in the distribution of grace by a necessarily just God who, Scripture says, wants everyone to be saved—is what most exercised the great philosophical minds of the period. This is most obviously true for such thinkers as Malebranche and Arnauld, but it is also true for someone with so wide a diversity of interests as Leibniz.[1] It is even true, in a sense, for Spinoza, although in a very different way: Rather than trying to solve the conundrum, he hoped to diminish the general concern with such questions by undermining the anthropomorphic and superstitious beliefs which nourished them. For Leibniz, Arnauld, and Malebranche, theodicy was a central element of their thinking; for Spinoza, it was a pastime for fools, and he wanted to make this perfectly clear.

Now a theodicy requires a conception of the deity whose justice and wisdom one is defending. It demands that one be committed to a particular way of conceiving the nature of God, and especially His attributes and His *modus operandi*. One's solution to the problem of evil—even one's determination as to whether or not the problem of evil can have a rational solution—will very much depend on just what one believes about the relationship between understanding, wisdom, will, and justice in God and on how most properly to describe the way in which God acts.

In the latter decades of the seventeenth century, it seemed to most thinkers that there were three models of God available for consumption. I do not want to defend the claim that these were in fact the only models available, but only that these were the conceptions of God that were most prominently represented—and debated—in the Republic of Letters on the Continent between 1677, when Spinoza died and

S. Nadler (✉)
University of Wisconsin-Madison, Madison, WI, USA
e-mail: smnadler@wisc.edu

[1]Rutherford (1995), for example, has argued for the primacy of philosophical theology, and especially the problem of evil, over other concerns (metaphysical and logical) in Leibniz.

C. Fraenkel et al. (eds.), *The Rationalists: Between Tradition and Innovation*, The New Synthese Historical Library 65, DOI 10.1007/978-90-481-9385-1_10,

his *Ethics* was finally published in Latin and Dutch editions, and 1716, when the aggregate of simple substances constituting Leibniz finally lost its dominant monad, 6 years after the publication of his *Theodicy*.

In this paper, after a relatively static examination of these three conceptions of God, I examine a certain dynamic between them. My primary aim, however, is not to provide a comprehensive investigation of the various ways of thinking about God in the period. Rather, I want to use these in order to consider what I find to be a very interesting passage. It comes from Spinoza's *Ethics*, and tells us something about what Spinoza thought about Leibniz's kind of God. I find the passage interesting especially because I find it puzzling, both for what is being said and for who is saying it.

10.1 Three Gods

I begin by sketching three philosophical conceptions of God in the seventeenth century: the rationalist God, the voluntarist God, and the God of Spinoza. The three conceptions have much in common. All attribute to God some basic essential characteristics: eternity, necessity, and infinitude, as well as being the ultimate causal power behind all things. But they are also distinguished by the different models of divine agency that inform them, by the different ways in which they portray God's *modus operandi*.

First, the rationalist God—by which I mean the conception of God that depicts God as acting in a rational manner, but not necessarily the God adopted by the most rationalist thinkers.

Odd as it may seem, an occasionalist like Malebranche and a concurrentist like Leibniz can share a conception of God. Both Malebranche and Leibniz are committed to the view that God is a rational being who does things for an intelligible and objective purpose. A rational being is one for whom reasons matter. Such an agent is motivated teleologically by aims; he acts for the sake of achieving something. And he strives to achieve what he does because he recognizes it as good, as desirable in its own right. Moreover, his rationality is instrumental: he selects means toward his desired goal because he believes, with justification, that those means are the most efficient way to it. To pursue ends that one does not believe to be good, or to follow a path toward one's end that one knows not to efficiently lead there, is to act irrationally.

Now consider Leibniz's God. He contemplates an infinite number of possible worlds and recognizes that one of them (the one that contains the simplest laws and the greatest amount of perfection) is, in absolute terms, the best of all. His desire is to produce as much perfection as possible, and so He brings that best of all possible worlds into existence. The optimality of that world provides Him with a compelling reason to create it; and had there been no such compelling reason, Leibniz's God would not have created a world at all. The principle of sufficient reason is binding even upon God. As Leibniz says, "[God] does nothing without acting in accordance with supreme reason" (*Theodicy*, §8, GP VI.107; H 128). This is true for all of

God's choices, large and small. It is beautifully reflected in Leibniz's argument for his famous law of the identity of indiscernables, which says that "there is no such thing as two individuals indiscernible from each other." The reason why there are not and cannot be two distinct things in nature that are absolutely identical in all intrinsic respects—two leaves, or two snowflakes, "without any difference within themselves"—is because, in the complete absence of any differences between the two, God would have no compelling reason to put one of them in one place and the other in another place, rather than vice versa. Consequently, God, "who never does anything without wisdom," will not create such things.[2]

God, then, on Leibniz's view, is never indeterminate; He never acts without knowledge of what He is doing and without being determined by reasons to do it. "His will is always decided, and it can be decided only by the best." God can never have what Leibniz calls a "primitive particular will," that is, an *ad hoc* volition that is independent of any law or principle. "Such a thing would be unreasonable. [God] cannot determine upon Adam, Peter, Judas, or any individual without the existence of a reason for this determination; and this reason leads of necessity to some general enunciation. The wise mind always acts according to principles; always according to rules, and never according to exceptions" (*Theodicy*, §337, GP VI.315; H 328).

Leibniz recognizes that this amounts to a kind of constraint on God, but insists that it is no more than the ordinary kind of determination that reasons bring to bear on rational choices. His critics, less sanguine about the consequences of putting restrictions on God than Leibniz was, were concerned that if these reasons lead God "of necessity," then God loses His freedom of choice. Pierre Bayle argued that Leibniz subjects God to a kind of fate: "There is therefore no freedom in God; He is compelled by His wisdom to create, and then to create precisely such a work, and finally to create it precisely in such ways. These are three servitudes that form a more than Stoic *fatum*, and that render impossible all that is not within their sphere."[3]

But the necessity that binds God, Leibniz says, is "a happy necessity," because it has its source in God's own nature. "This so-called *fatum*, which binds even the Divinity, is nothing but God's own nature, his own understanding, which furnishes the rules for his wisdom and his goodness" (*Theodicy*, §191, GP VI.230; H 246–7). Contrary to Bayle's accusation, other possible worlds still remain possible, logically speaking. Each, considered in and of itself, involves no logical contradiction (as would be the case, for example, with a world in which both Napoleon is the French emperor and Napoleon is not the French emperor). While these possibilities will, with certainty, *not* be chosen by God, and thus do not remain live options for

[2]See Leibniz's fourth letter to Samuel Clarke, GP VII.372; L 687. This is only one of many occasions where Leibniz employs this argument for his law. See also Leibniz's Fifth Paper (for Clarke): "I infer from [the principle of sufficient reason] ... that there are not in nature two real, absolute beings, indiscernible from each other, because if there were, God and nature would act without reason in treating the one otherwise than the other; and that therefore God does not produce two pieces of matter perfectly equal and alike" (GP VII.393; L 699 [translation modified]).

[3]Quoted by Leibniz at *Theodicy*, §227, GP VI.253; H 268.

His creative will, nonetheless their conceptual reality is sufficient to guarantee that God's determined choice of the best world is a free choice. The determination is characterized only by what Leibniz describes as a *moral* (as opposed to a logical, metaphysical, or absolute) necessity, the necessity of reasons that compel choice in a rational being—a being who, because of His goodness and wisdom, is moved infallibly to choose the best. "There is always a prevailing reason that prompts the will to its choice… The choice is free and independent of [absolute] necessity because it is made between several possibles, and the will is determined only by the prepondering goodness of the object" (*Theodicy*, §45, GP VI.1127–8; H 148).

In his early works, Malebranche often seems more interested in defending the sheer power of God's will than its wisdom and rationality. "God is a being whose will is power and infinite power," he wrote in the *Christian Conversations* (OC IV.30–31).[4] But even in the *Search After Truth*, and especially in the *Treatise on Nature and Grace*, it is clear that Malebranche's God—like Leibniz's God—does not act arbitrarily. This is true in the ordinary course of nature (and grace), where God's causal activity is guided by the laws. But the fact that God must choose the most simple laws for the world He creates, and then strive to produce as just and perfect a world as possible relative to those laws, shows that there is a higher authority than His will alone. God, Malebranche says, "cannot act against Himself, against His own wisdom and light." He may be indifferent as to whether or not anything other than Himself exists, that is, indifferent about whether or not to create at all. But having decided to create, "[God] is not indifferent, although perfectly free, in the way in which He does it; He always acts in the wisest and most perfect way possible. He always follows the immutable and necessary order" (*Search After Truth*, Elucidation 8, OC III.85–6; LO 586–7).

God's power is in itself infinite and incomprehensible. However, once Malebranche's God does decide to act "external to Himself," that power is subordinated to His wisdom, and especially to what Malebranche calls Order. Order consists in the eternal, immutable verities that stand above all things. These are pure logical and mathematical truths, absolutely true with the highest necessity, but also moral and metaphysical principles about what Malebranche calls "relations of perfection." They determine the relative value of various kinds of being, and even of God's own attributes. Order shows that a soul is more noble than a body, and a human being more worthy than a dog; and it proclaims that, as important as God's mercy is, the simplicity and generality of His ways are even more important, and thus cannot be violated even to save a good person from drowning or damnation.[5] Order is "the exemplar of all God's works," and his volitions must conform to its principles. The dictates of Order serve God as universal reasons for everything He does. Even if, on some rare occasion, God must act by a particular volition and violate the laws of nature or grace to perform a miracle, this will be only because Order demands it (TNG, II.45).

---

[4]For a discussion of this, see André Robinet (1965, 22).

[5]See *Dialogues on Metaphysics*, VIII.13.

Thus, when God, considering the infinite possibilities in His understanding, chooses to create a world, Order sets one of His attributes (simplicity) above the others, which in turn determines which laws He will establish for the world, and then how, given those laws, He can thereby accomplish as much perfection as possible in the relationship of the physical to the moral, that is, of the relationship between the pain of punishment or the pleasures of reward, on the one hand, and, on the other hand, sin or virtue.[6]

God's wisdom, the dwelling place of Order, stands above his will and guides it. "In Himself," Malebranche says, "God has good reasons for everything He does" (*Dialogues on Metaphysics*, IX.3, OC XII.201; JS 152).

Like Leibniz, Malebranche is concerned that his account of divine rationality might be seen as compromising God's freedom. He even concedes that, in a manner of speaking, it does. "The wisdom of God renders Him impotent in the following sense, that it does not allow Him to will certain things, nor to act in certain ways ... God is impotent in the sense that he cannot choose ways of acting unworthy of his wisdom, or that do not bear the character of his goodness, of his immutability, or of his other attributes" (TNG, Third Elucidation, OC V.180). But, once again, this is only the *moral* impotence of a perfectly rational being to act contrary to unassailable reasons. God's choices are compelled or determined, there are standards that God is bound to observe. But these standards lie in God's wisdom, and the obligation to obey them comes from His nature alone.

Thus, the rationalist God. By contrast, consider the voluntarist God of Descartes and Arnauld, particularly as this is expressed in the Cartesian doctrine of the creation of the eternal truths. Descartes, famously siding with Euthyphro, insisted that "if anyone attends to the immeasurable greatness of God, he will find it manifestly clear that there can be nothing whatsoever which does not depend on Him. This applies not just to everything that subsists, but all order, every law, and every reason for anything's being true or good" (Sixth Replies, AT VII.435; CSM II.293–4). God is therefore absolutely "indifferent" and undetermined not only with respect to whether or not to create a world and whether to create this world rather than any other, but also with respect to what truths and laws He establishes.[7]

> God did not will the creation of the world in time because He saw that it would be better this way than if He had created it from eternity; nor did He will that the three angles of a triangle should be equal to two right angles because He recognized that it could not be otherwise, and so on. On the contrary, it is because He willed to create the world in time that it is better this way than if He had created it from eternity; and it is because He willed that the three

---

[6]"Assuming that God wants to act, I contend that he will always do it in the most wise manner possible, or in the manner which most bears the character of his attributes. I insist that this is never an arbitrary or indifferent matter for Him ... Immutable Order, which consists in the necessary relationship that exists between the divine perfections, is the inviolable law and the rule of all His volitions" (Reply to Arnauld's *Philosophical and Theological Reflections on the New System of Nature and Grace*, OC VIII.752–3).

[7]For the relevant texts, see note 2 above. Jean-Marie Beyssade (1979, chap. 2) argues that indifference is a feature of the doctrine that is absent from the 1630 exchange and informs only the 1644 letters.

> angles of a triangle should necessarily equal two right angles that this is true and cannot be otherwise; and so on in other cases. (Sixth Replies, AT VII.435; CSM II.293–4)

Descartes goes on to say that "if some reason for something's being good had existed prior to His preordination, this would have determined God to prefer those things which it was best to do. But on the contrary, just because He resolved to prefer those things which are now to be done, for this very reason, in the words of Genesis, 'they are very good'; in other words, the reason for their goodness depends on the fact that He exercised His will to make them so" (Sixth Replies, AT VII.435–6; CSM II.294).

The assumption behind this doctrine is that, in an absolutely simple and omnipotent being such as God, will and understanding are one and the same thing, distinguishable *ne quidem ratione*. "In God, willing and knowing are a single thing in such a way that by the very fact of willing something He knows it and it is only for this reason that such a thing is true" (Letter to Mersenne, May 6, 1630, AT I.149; CSMK III.24). Thus, for God to know that 1+1=2 is identical to God willing that 1+1=2. Descartes's point is that the relationship between will and wisdom in God is so unlike their relationship in human beings—where will and understanding *are* distinct faculties of the mind, and the former cannot function unless the latter presents to it ideas for consideration[8]—that no analogy can be drawn from the human for the purpose of understanding the divine. As Jean-Luc Marion has so eloquently shown, what Descartes is rejecting is the Scholastic (Suarezian) doctrine of the univocity of understanding between God and creatures, whereby terms employed for the conception of the former's mode of understanding are applicable in a univocal (or even analogical) sense to the conception of the latter's mode of understanding (Marion 1981). Nothing, for Descartes, "can belong *univoce* to God and to creatures" (Sixth Replies, AT VII.433; CSM II.292). God's activity, then, is not to be modeled on practical rational agency, lest we "conceive of God as if He were some kind of superman, who sets this or that end for himself, and strives for it by these or those means, which is certainly completely unworthy of God" (Conversation with Burman, AT V.158).

In Arnauld, this model finds perhaps an even more vehement defense. Part of Malebranche's problem according to Arnauld is that to distinguish wisdom from will in God—even by a "distinction of reason," grounded not in reality but in the way things are conceived—and have wisdom guide will by providing compelling reasons for its choices is to undermine divine freedom. Malebranche repeatedly says that "God's wisdom renders Him, in a sense, impotent" by determining Him to choose one world rather than another.[9] Malebranche takes comfort in the "in a sense" qualification, as well as in God's original indifference as to whether or not to create a world in the first place, and so is not particularly troubled by the implications of this for His freedom. Arnauld, however, is. He conceives of God's

[8] This process is described, for example, in the Fourth Meditation.

[9] In addition to the passage cited above, see TNG, OC V.180, 185.

freedom as consisting in an absolute "liberty of indifference," thoroughly undetermined in the creation and governance of things. God's will is not guided by anything whatsoever external to the will itself, not even by the dictates of His own wisdom.[10]

Malebranche's God, Arnauld claims, cannot possibly satisfy what he sees as Aquinas's authoritative demand that the will of God remain perfectly self-determining, never willing anything external to itself *ex necessitate* (*Reflections*, OA XXXIX.598–99). To be fair, Malebranche, despite his deterministic language, strives to preserve the ultimate contingency of God's creative act. But—and this is Arnauld's point—Malebranche's account fails miserably; he ends up subjecting God to "a more than stoical necessity."[11] In fact, Arnauld appears to be saying, how could it be otherwise? In a perfectly rational being, in whom there are no passions exercising a contrary influence, reasons would have to determine and necessitate the will and render it "impotent" to choose otherwise. When His wisdom dictates the creation of one world over all the others, Malebranche's God's choice is thereby infallibly determined; He *must* create that world, Arnauld insists (and Malebranche apparently agrees). Because they are demanded by Order, not even miracles are freely ordained by Malebranche's God (*Reflections*, OA XXXIX.599). Thus, in the name of divine freedom, Arnauld rejects the thesis, popular among medieval philosophers and now adapted by Malebranche to his own theodicean purposes, of the *priority* of divine understanding over divine will. In Arnauld's God, *sagesse* does not rule over *volonté*.

Thus, Arnauld warns that Malebranche's distinction between wisdom and will in God and his rationalistic depiction of God's behavior constitute an anthropomorphizing of God's nature and of His ways. It portrays God as if He, like human beings, has a mind constituted by different faculties, with a will that is able to select volitionally only from among the options that the understanding presents to it and that is guided in its choice by the dictates of reason:

> [Malebranche] speaks about [God] as if he were speaking about a human, in making Him consult His wisdom on everything He would like to do . . . as if His will, in order to will nothing other than what is good, had need of being regulated by something other than itself." (*Reflections* II.26, OA XXXIX.599–600)

---

[10] *Reflections philosophiques et théologiques sur le nouveau système de la nature et de la grace*, OA XXXIX.600. According to Arnauld, it also generates a problem of consistency for Malebranche because Malebranche does want to say that God *is* indifferent in the initial choice to create a world outside Himself.

[11] "By following Malebranche in the manner in which he conceives God, I do not see how He can be indifferent to creating or not creating something outside Himself, if He was not indifferent to choosing among several works and among several ways of producing them. For God . . . , according to [Malebranche], having consulted His wisdom, is necessarily determined to produce the work that it [wisdom] has shown him to be the most perfect, and to choose the means that it has shown Him also to be the most worthy of Him" (*Reflections*, OA XXXIX.599). As Robert C. Sleigh, Jr., points out (1990, 45–47), this concern (worded in almost exactly the same way) reappears less than 20 years later in his criticisms of Leibniz.

Arnauld, like Descartes, sees God as a being in whom will and wisdom are one and the same, and thus for whom the will is a law unto itself. This God indifferently creates reasons through its volitions. He does not, like Malebranche's God, have a will that takes its lead from wisdom's logically antecedent reasons.[12] Arnauld's God is a deity who does not act for reasons at all, a deity who, in the structure of His being, transcends practical rationality altogether.

Finally, there is the God of Spinoza. Spinoza denies the traditional God of Abrahamic religions as a pernicious anthropomorphic fiction. God, he insists, is not endowed with any of the psychological or moral characteristics necessary for a providential Being. God does not will, judge, command, or deliberate. Nor is God good, wise, and just. Spinoza's God is not a personal God to whom one would pray or seek solace in times of trouble; it is not a God that can be the object of reverential awe or worshipful submission. For Spinoza, God is nothing but nature—*Deus sive Natura*, "God or Nature," in his famous phrase—the infinite, necessarily existing, eternal, and active substance of the universe. There is nothing good or bad, perfect or imperfect about God or Nature; it just *is*.

Spinoza's God does not choose the best of all possible worlds. Spinoza's God, in fact, does not choose anything whatsoever. Spinoza's identification of God with the eternal, infinite, necessarily existing substance of Nature itself—with the most general natures of things (Thought and Extension) and the universal causal principles and laws embedded in these—means that whatever exists within Nature (and this is everything possible) "follows from" or is caused by God or Nature with an absolute, inevitable necessity. Nothing whatsoever could possibly have been otherwise: not the universe itself, nor any individual thing or event within it. "All things, I say, are in God, and all things that happen, happen only through the laws of God's infinite nature and follow ... from the necessity of his essence" (*Ethics* IP15S[VI], G II.60; C 424).

The metaphysics of God in the *Ethics*, motivated as it is by an extreme anti-anthropomorphism, rules out any depiction of God that involves Him considering alternative possibilities, acting for purposes, making choices based on reasons, and assessing outcomes. "There are those who feign God, like man, consisting of a body and a mind, and subject to passions. But how far they wander from the true knowledge of God, is sufficiently established by what has already been demonstrated" (*Ethics* IP15S[I], G II.57; C 421). All talk of God's purposes, intentions, goals, preferences, or aims is just a fiction propagated by manipulative ecclesiastics. "All the prejudices I here undertake to expose depend on this one: that men commonly

---

[12]"If we are asked why God has created the world, we should reply only that it is because He wanted to; and ... if we are asked again why He wanted to, we should not say, as [Malebranche] does, that 'He wanted to obtain an honor worthy of Himself.' The idea of God does not allow us to accept Malebranche's proposition. We ought rather to say that He wanted to because He wanted to, that is, that we ought not to seek a cause of that which cannot have one" (*Reflections* II.3, OA XXXIX.433). Vincent Carraud (1996, 91–110) notes that this is Arnauld's refusal "to submit God to causality, that is, to submit His will to rationality in the form of a principle of reason." See also Thomas M. Lennon (1978, 186); Lennon recognizes that, contrary to Malebranche's "rationally constrained" God, for Arnauld "divine self-determination [is] utterly unconstrained and thus mysterious."

suppose that all natural things act, as men do, on account of an end; indeed, they maintain as certain that God himself directs all things to some certain end, for they say that God has made all things for man, and man that he might worship God" (*Ethics* I, Appendix, G II.78; C 439–40).

Spinoza's God, then, unlike the rationalist God, is not some goal-oriented planner who then judges things by how well they conform to his purposes. Things happen only because of Nature and its laws. "Nature has no end set before it . . . All things proceed by a certain eternal necessity of nature." To believe otherwise is to fall prey to the same superstitions that lie at the heart of the organized religions.[13] A rationalist, judging God who has plans and acts purposively is a God to be obeyed and placated. Opportunistic preachers are then able to play on our hopes and fears in the face of such a God. They prescribe ways of acting that are calculated to avoid being punished by that deity and earn His rewards. But, Spinoza insists, to see God or Nature as acting for the sake of ends—to find purpose in Nature—is to misconstrue Nature, to "turn it upside down" by putting the effect (the end result) before the true cause. In Spinoza's view, the traditional religious conception of God leads only to superstition, not enlightenment.

In the end, Spinoza's God is neither an arbitrary nor a rational deity. The existence of God or Nature itself is absolutely necessary—it cannot not exist; and whatever happens in nature—everything that has been or will be—is caused by God or Nature, not by choice but by necessity, and thus comes about through natural principles with a geometric inevitability. For Spinoza, this is not the best of all possible worlds; it is the only possible world.

10.2 Spinoza's Choice

So much for a static picture of the three Gods of philosophy in the seventeenth century. The dynamics between the three are even more interesting. There are, as we have seen, the criticisms that Arnauld levels against the rationalist God, criticisms

---

[13]"[People] find—both in themselves and outside themselves—many means that are very helpful in seeking their own advantage, e.g., eyes for seeing, teeth for chewing, plants and animals for food, the sun for light, the sea for supporting fish. Hence, they consider all natural things as means to their own advantage. And knowing that they had found these means, not provided them for themselves, they had reason to believe that there was someone else who had prepared those means for their use. For after they considered things as means, they could not believe that the things had made themselves; but from the means they were accustomed to prepare for themselves, they had to infer that there was a ruler, or a number of rulers of nature, endowed with human freedom, who had taken care of all things for them, and made all things for their use. And because they had never heard anything about the temperament of these rulers, they had to judge it from their own. Hence, they maintained that the gods direct all things for the use of men in order to bind men to them and be held by men in the highest honor. So it has happened that each of them has thought up from his own temperament different ways of worshipping God, so that God might love them above all the rest, and direct the whole of Nature according to the needs of their blind desire and insatiable greed. Thus this prejudice was changed into superstition, and struck deep roots in their minds" (*Ethics* I, Appendix, G II.78–9; C 440–41).

made in the name of defending divine omnipotence and freedom. And there are the criticisms that Leibniz and Malebranche both make of the voluntarist God, criticisms made in the name of defending divine wisdom, goodness, and justice. Leibniz and Malebranche insist that if God is as Descartes says He is, then claims about God's wisdom and justice are trivial or meaningless.

Leibniz was a lifelong foe of divine voluntarism.[14] In his mind, the flaws of Descartes's doctrine run deep. Not only does it offer a false picture of God's nature and of the way in which He operates, but it threatens religious piety and even everyday morality. Above all, he believes that the moral character of God Himself is threatened by the voluntarist position.

> In saying, therefore, that things are not good according to any standard of goodness, but simply by the will of God, it seems to me that one destroys, without realizing it, all the love of God and all His glory; for why praise Him for what He has done if He would be equally praiseworthy in doing the contrary? Where will be His justice and His wisdom if He has only a certain despotic power, if arbitrary will takes the place of reasonableness, and if in accord with the definition of tyrants, justice consists in that which is pleasing to the most powerful?" (*Discourse on Metaphysics* §2, GP IV.427–8; M 4–5)[15]

Malebranche makes a similar claim. If Descartes is right, he says, then claims about God's moral nature are empty and there is no point in praising God for what He has done because He would be equally praiseworthy had He done just the opposite.[16]

There was, however, one point on which all hands agreed—whether they were partisans of the rationalist God or partisans of the voluntarist God. It was that if there was anything to be avoided, it was the God of Spinoza. Leibniz, Malebranche, and Arnauld all spent a good deal of time either trying to clear themselves of the charge of Spinozism or accusing others of holding views with Spinozist implications. Everyone was trying to avoid the Spinozistic vortex.

---

[14]Marion (1981, 181–296) rightly cautions against using the label 'voluntarism' for Descartes's position, insofar as it suggests a continued distinction between divine faculties, except now with will having priority over understanding—which is incompatible with Descartes's insistence that there is no such distinction in God. However, the term, while misleading in this way, does serve well to highlight the fact that for Descartes, the eternal truths are dependent on God's causal power, even if His willing those truths is identical with His understanding them.

[15]Marion (1985, 143–164) argues that in formulating his critique of Descartes's doctrine, Leibniz willfully misreads it to introduce a priority of will and understanding in God.

[16]"According to this principle, the universe is perfect because God willed it. Monsters are works as perfect as others according to the plans of God. It is good to have eyes in our head, but they would have been as wisely placed anywhere else, had God so placed them. However we invert the world, whatever chaos we make out of it, it will always be equally admirable, since its entire beauty consists in its conformity with the divine will, which is not obliged to conform to order. All the beauty of the universe must therefore disappear in view of that great principle that God is above the Reason that enlightens all minds, and that His wholly arbitrary will is the sole rule of His actions" (*Dialogues on Metaphysics* IX.13, OC XII.220–221; JS 168–69).

And this brings us, at last, to the passage that I want to examine. For it constitutes a very interesting but somewhat perplexing moment in the dynamic among the three conceptions of God.

The passage comes from the second scholium to proposition thirty-three of Part One of the *Ethics*. IP33 itself is the proposition in which Spinoza addresses the idea that God could have made things differently—basically, not just that there are other possible worlds, but that they are possible worlds that God could have instituted. Spinoza says that "things could have been produced by God in no other way, and in no other order than they have been produced." With this proposition, of course, Spinoza is essentially denying a thesis that is at the heart of both the rationalist God of Leibniz and the voluntarist God of Descartes—both conceptions allow that there is a sense in which God could have chosen otherwise, although they allow this in very different ways (with Leibniz actually coming close to denying it). But, at the end of the second scholium to this proposition, Spinoza goes on to say that if forced to choose between the rationalist God and the voluntarist God, he would choose the voluntarist God.

> I confess that this opinion, which subjects all things to a certain indifferent will of God, and makes all things depend on his good pleasure, is nearer the truth than that of those who maintain that God does all things for the sake of the good. For they [the latter] seem to place something outside of God, which does not depend on God, to which god attends as a model, in what he does, and at which he aims, as at a certain goal. This is simply to subject God to fate. Nothing more absurd can be maintained about God, whom we have shown to be the first and only free cause, both of the essence of all things, and of their existence. So I shall waste no time in refuting this absurdity.

On the face of it, what Spinoza is saying here might seem perfectly clear. The rationalist God conforms to some external standards of truth and goodness, and these objective and independent normative constraints represent limitations on God's power. And this is inconsistent with the idea that, because God is the sole substance of Nature, there is nothing that is external to and independent of God; it is also inconsistent with Spinoza's conception of God's absolute freedom, understood as the claims embodied in IP17: "God acts from the laws of his nature alone, and is compelled by no one.... There can be nothing outside him by which he is determined or compelled to act. Therefore, God acts from the laws of his nature alone"; and in the second corollary to IP17: "God alone is a free cause. For God alone exists only from the necessity of his nature and acts from the necessity of his nature." The rationalist model, in Spinoza's eyes, threatens divine power and freedom by subjecting God to some binding external norms, whereas the voluntarist model leaves God perfectly self-determining, the spontaneous active cause on which everything whatsoever depends—something more resembling the free causality of Spinoza's God.

However, what I find odd about Spinoza's comments here is that they are contrary to what the reader of the *Ethics* might reasonably expect. One would think that it is not the voluntarist, Cartesian God that Spinoza should favor, but the rationalist God—Leibniz's God! Of course, Spinoza cannot possibly accept the character of

Leibniz's God.[17] Spinoza's God does not have as part of its essence either will or intellect; and Spinoza's God is certainly not capable of rational choice. However, why should this count in favor of the voluntarist God? The Cartesian God really fares no better in this regard; although there is no distinction between will and understanding in the Cartesian God, He is still a volitional being making absolute choices. Moreover, the Cartesian God embodies a conception of divine freedom grounded in "an absolute will" making arbitrary choices that could just as easily have been different, something that Spinoza explicitly labels "absurd" and regards as something to be rejected "not only as futile, but as a great obstacle to science" (IP33s2).

So again, why does Spinoza prefer the voluntarist God? Part of the puzzlement generated by the passage is that Spinoza's critique of the rationalist God seems a little misguided. Leibniz's God does not have to answer to moral and epistemic values that are "outside" God, that "do not depend on God." The goodness and truth of what is good and true are indeed, for Leibniz and his predecessors in this matter, independent of God's will. But, like the eternal truths in Augustine and Aquinas and like Malebranche's Order, they are not absolutely independent of and external to God insofar as they are embedded in God's wisdom. In choosing the best of all possible worlds, the rationalist God—either Leibniz's or Malebranche's—is not complying with any *external* standards. Rather, He is only following through on the determinations of His own nature and on the objective normative demands set among the possibilities that He finds in His own understanding. As Leibniz says in the second article of the *Discourse on Metaphysics*, while all acts of God's will "presuppose a reason for willing and that this reason is naturally prior to the act of will," it is also the case the reasons that move God are not external to Him but "consequences of His understanding" that do not "depend on His will" (GP IV.428; M 5). Oddly, then, if Spinoza's complaint in the passage is to make sense, it must be the objection that a God like Leibniz's God is a God for whom the will is determined by something outside the will—which is true, but it is a strange complaint to be made by someone who insists that God does not have will in the first place.

Thus, given Spinoza's own claim, in the first corollary to IP32, that God does not act with an absolute and indifferent freedom of will; and his consequent denial of the absolute contingency of all things, Spinoza really should prefer the rationalist God to the voluntarist God. For the rationalist God can quite conveniently accommodate Spinoza's insistence in IP33 that "things could have been produced by God in no other way, and in no other order than they have been produced." Indeed, Leibniz himself, always sensitive to the charge of Spinozism, nonetheless concedes that, once God's nature and choice is taken into account, then the existence of the actual world (as the best of all possible worlds) is necessary, that God could not have produced any of the other possible worlds. Commenting on IP33 of the *Ethics*, for

[17] I refer here to Spinoza confronting Leibniz's God, although of course Spinoza knew nothing about Leibniz's own conception of God at least until their meeting in 1676, over a decade after he was probably finished composing Part One of the *Ethics*. What Spinoza is referring to in the passage is the traditional, rationalist conception of God found among medieval philosophers such as Maimonides and Aquinas.

example, Leibniz replies that "on the hypothesis that the divine will chooses the best or works in the most perfect way, certainly only this world could have been produced" (A VI.4b.1776; L 204).

Why, then, does Spinoza *not* prefer the rationalist God to the voluntarist Cartesian God? I suppose that the explanation offered above that relies on Spinoza's concern to defend a proper conception of divine activity, power, and freedom—that is, to safeguard God or Nature's causal autonomy and spontaneity from being encumbered and determined by "external" standards—can answer this, although as I suggest, it does seem to rely on a misreading of the relationship between the rationalist God and the values that inform His choices.

But here is another suggestion. Matthew Stewart (2006) has argued that Leibniz was deeply fascinated, even obsessed by Spinoza's philosophy, and that most of his own thought is developed with one eye on the Spinozist corner. I do think that Stewart is over-stating things a bit. But what I want to emphasize here is that it can be said that Spinoza, too, is very concerned with keeping a particular conception of God at a distance—in fact, Leibniz's type of God. And this, I suggest, is what is behind the passage in question. Despite the reasons that Spinoza explicitly offers in the passage for his hypothetical preference, it is not just divine freedom from external constraint that is at stake, but anthropomorphism and its concomitant conception of divine providence. Descartes's God, in whom there is no distinction between will and understanding, is simply less anthropomorphic a deity than the rationalist God, who, like us, acts according to practical reason.[18]

That this is a concern of Spinoza's is already clear in the *Metaphysical Thoughts* appended to his early *Descartes's Principles of Philosophy*, in which he says that

> God's will, through which He has created things, is not distinct from His intellect, through which He understands them. So to say that God understands that the three angles of a triangle are equal to two right angles is the same as saying that God has will or decreed that the three angles of a triangle should equal two right angles. (II.4, G 1.257; C 322–3)

He concludes that those who have "attributed a human will to [God], i.e., a will really distinct from the intellect" have "no knowledge at all of the nature of God's will" (II.7, G 1.261; C 326). By the *Ethics*, as we have seen, will and understanding no longer pertain to God's essence at all. Moreover, Spinoza insists there that even if intellect and will *did* pertain to the essence of God,

> we must understand by each of these attributes something different from what men commonly understand. For the intellect and will which would constitute God's essence would have to differ entirely from our intellect and will, and could not agree with them in anything except the name. They would not agree with one another any more than do the dog that is a heavenly constellation and the dog that is a barking animal. (IP17s2)

And of course, the relationship between will and intellect that Spinoza attributes to human beings is quite different from the way in which it is depicted by Descartes. Spinoza famously insists that these are not two distinct faculties in us; rather,

---

[18]This is especially so given the way "men commonly," including Descartes (but not Spinoza), conceive of the relationship between will and understanding in human beings.

so-called acts of will belong to ideas themselves—that is, every idea of the human understanding essentially involves a volitional component.[19] So to anthropomorphize God and attribute to Him a will and understanding as these are *truly* conceived in human beings—that is, as Spinoza conceives them—does not necessarily culminate in a God in whom intellect and will are distinct faculties. It would, however, still have God acting the way we act.

As Leibniz himself insists, God must be a kind of "person," and this is what Spinoza cannot possibly tolerate.[20] Such a God involves deliberate choice over a range of possibilities, and thus is a God who exercises providential care for His creation in a strong sense. If Part One of the *Ethics* is about anything, it is about eliminating anthropomorphic elements from the true philosophical conception of God's nature and correcting a traditional but false understanding of divine providence—in a way, carrying through Maimonides' project to its ultimate logical conclusion. This is, perhaps, why, in the passage, Spinoza chooses what initially seems *not* to be the lesser of two evils.

I think Leibniz himself saw all this very clearly. Consider what he says in his letter to Gerhard Molanus from around 1679. After offering his standard criticism that "Descartes's God, or perfect being, is not a God like the one we imagine or hope for, that is, a God just and wise, doing everything possible for the good of creatures," Leibniz adds that

> Descartes's God is something approaching the God of Spinoza, that is, the principle of things and a certain sovereign power or primitive Nature that puts everything into action and does everything doable. Descartes's God has neither will nor understanding, since according to Descartes He does not have the good as the object of the will, nor the true as the object of the understanding. (GP IV.299; AG 242)

Again, it seems on the face of it to be an odd claim. Leibniz is insisting that the Cartesian voluntarist God whose will is independent of all determining reasons closely resembles the Spinozist deity. It is odd because, while Descartes's God is endowed with an absolute will making an undetermined choice, Spinoza's God wills nothing at all.

But Leibniz's point here is that Descartes, by eliminating *rational* choice in God, essentially eliminates divine choice altogether. True choice for Leibniz is necessarily an informed choice among alternative possibilities guided, even determined by objective reasons—it is the choice of a moral agent who, in the absence of compelling reasons, simply does not choose; whereas an arbitrary volition that could just as easily have been directed at the opposite is neither rational nor characteristic of moral agency, and thus not a true choice at all. Thus, a God who, like Descartes's God, does not will for the sake of reasons and to achieve what is good is ultimately not really endowed with intellect and will at all—just like the God of the *Ethics*. Indeed, just like Spinoza's God, it is not a God who is, as Leibniz demands, a "person."

---

[19] See IIP49 and its scholia.

[20] See A VI.3.474–5.

So we have, in the end, found here, through an initially paradoxical passage, a point on which both Spinoza and Leibniz agree: that Spinoza's God has much more in common with the Cartesian God than with Leibniz's God. Spinoza's God and the Cartesian God lack the moral and psychological—that is, anthropomorphic—features that are essential for the exercise of providence in the strong sense. In Leibniz's eyes, this represents their shortcoming. In Spinoza's eyes, it is their virtue.[21]

Abbreviations

Works by Leibniz

A *Sämtliche Schriften und Briefe*. Edited by Deutsche Akademie der Wissenschaften. Multiple vols. in 7 series. Darmstadt/Leipzig/Berflin: Akademie Verlag, 1923.

AG *Philosophical Essays*. Edited and translated by Roger Ariew and Daniel Garber. Indianapolis: Hackett, 1989.

GP *Philosophische Schriften*. Edited by C. I. Gerhardt. 7 vols. Berlin: Weidmann, 1875–90. Reprint, Hildesheim: Georg Olms, 1978.

H *Theodicy: Essays on the Goodness of God, the Freedom of Man, and the Origin of Evil*. Edited by Austin Farrar. Translated by E. M. Huggard. La Salle, IL: Open Court, 1985.

L *Philosophical Papers and Letters*. Edited and translated by Leroy Loemker, 2nd. ed. Dordrecht: D. Reidel, 1969.

M *Discourse on Metaphysics / Correspondence with Arnauld / Monadology*. Edited and translated by George Montgomery. La Salle, IL: Open Court, 1980.

Works by Malebranche

JS *Dialogues on Metaphysics and on Religion*. Edited by Nicholas Jolley. Translated by David Scott. Cambridge: Cambridge University Press, 1997.

LO *The Search After Truth*. Translated by Thomas M. Lennon and Paul J. Olscamp. Columbus: Ohio State University Press, 1980.

OC *Oeuvres complètes de Malebranche*. Edited by André Robinet. 20 vols. Paris: J. Vrin, 1958–76.

TNG *Traité de la nature et de la grace*.

[21] For their helpful feedback on this paper and on the issues addressed, I am very grateful to the participants in the conference on Leibniz and Spinoza held at Princeton University in September 2007, members of the Philosophy Faculty at the University of Amsterdam in June 2007, colloquium audiences in the philosophy departments at Notre Dame and Indiana University in October 2007, and especially colleagues in the Montreal Interuniversity Workshop in the History of Philosophy in November 2006.

Works by Arnauld

OA *Oeuvres de Messire Antoine Arnauld, docteur de la maison et société de Sorbonne*. 43 vols. Lausanne: Sigismond d'Arnay, 1775.

Works by Descartes

AT *Oeuvres de Descartes*. 12 vols. Edited by Charles Adam and Paul Tannery. Paris: J. Vrin, 1974–83.

CSM *The Philosophical Writings of Descartes*. 2 vols. Edited and translated by John Cottingham, Robert Stoothoff, and Dugald Murdoch. Cambridge: Cambridge University Press, 1985.

CSMK *The Philosophical Writings of Descartes*. Vol. 3 (correspondence). Edited and translated by John Cottingham, Robert Stoothoff, Dugald Murdoch, and Anthony Kenny. Cambridge: Cambridge University Press, 1991.

Works by Spinoza

G *Spinoza Opera*. 4 vols. Edited by Carl Gebhardt. Heidelberg: Carl Winters Universitätsverlag, 1925.

C *The Collected Works of Spinoza*. Vol. 1. Translated by Edwin Curley. Princeton: Princeton University Press, 1985.

References

Beyssade, Jean-Marie. 1979. *La philosophie première de Descartes*. Paris: Flammarion.

Carraud, Vincent. 1996. Arnauld: A Cartesian theologian? Omnipotence, freedom of indifference, and the creation of the eternal truths. In *Interpreting Arnauld*, ed. Elmar Kremer. Toronto, ON: University of Toronto Press.

Lennon, Thomas M. 1978. Occasionalism, Jansenism, and skepticism: Divine providence and the order of grace. *Irish Theological Quarterly* 45: 185–90.

Marion, Jean-Luc. 1981. *Sur la théologie blanche de Descartes*. Paris: Presses Universitaires de France.

Marion, Jean-Luc. 1985. De la création des vérités éternelles au principe de la raison. Remarques sur l'anti-cartésianisme de Spinoza, Malebranche, Leibniz. *XVIIe siècle* 37: 143–64.

Robinet, André. 1965. *Système et existence dans l'oeuvre de Malebranche*. Paris: J. Vrin.

Rutherford, Don. 1995. *Leibniz and the rational order of nature*. Cambridge: Cambridge University Press.

Sleigh, Robert C. Jr. 1990. *Leibniz and Arnauld: A commentary on their correspondence*. New Haven, CT: Yale University Press.

Stewart, Matthew. 2006. *The courtier and the heretic*. New York: Norton.

Chapter 11
Leibniz on Infinite Beings and Non-beings

Ohad Nachtomy

11.1 Introduction

The notion of being has played a central role in philosophy ever since its early days. Against Plato's notion of being, Aristotle has used the notion of substance to account for changes and variations of one and the same thing. The notion of substance as referring to being is especially central for thinkers from the rationalist school, such as Descartes, Spinoza, and Leibniz. Likewise, the rationalists' notion of substance suffered severe attacks from empiricist thinkers, such as Locke, Berkeley and Hume. Famously, Locke mocked the rationalists' notion of Substance as a mere pompous word for "I know not what," resembling children's use of words without having an idea what they mean by it. Hume famously denied a continuous and subsistent notion of substance beyond sequences of distinct perceptions, arguing that any relations between these perceptions must derive from our own mind. And Berkeley went as far as arguing that the notion of material substance is utterly contradictory.

While all this is very familiar to readers in modern philosophy, it seems to me that there is a neglected aspect in the rationalist's notion of substance that calls for some close attention. What I have in mind is the strong, if under-investigated, connection between substantiality and infinity. For all rationalists, God is an infinite and most perfect substance. According to Descartes, God creates two types of *finite* substance—thinking and extended. According to Spinoza, there is nothing but a unique and infinite Substance, God or Nature, which has infinitely many attributes, each of which has both finite and infinite modes. According to Leibniz, God creates a world (among infinitely many possible worlds) and each world consists of infinitely many substances, and each substance itself consists of infinitely many other substances.

It can be seen from these few remarks that a number of intriguing questions arise when one examines the relation between substantiality and infinity in the context of these rationalist thinkers. Indeed, this is a field of research to which I will dedicate a

O. Nachtomy (✉)
Bar-Ilan University, Ramat Gan, Israel; Fordham University, New York, NY, USA
e-mail: ohnadnachtomy@mac.com

C. Fraenkel et al. (eds.), *The Rationalists: Between Tradition and Innovation*,
The New Synthese Historical Library 65, DOI 10.1007/978-90-481-9385-1_11,

book length project. The present article is something of a chapter within this broader project and its scope is likewise much narrower. It focuses on Leibniz's distinctions between beings and non beings and their inherent relation to infinity.

I will examine various contexts in which Leibniz distinguishes beings from non-beings and various ways in which he articulates this distinction. More specifically, I focus on one of Leibniz's defining features of true beings, namely that they are infinite. I examine a number of contexts in which the notion of infinity plays a significant role. This examination reveals that, according to Leibniz, infinity is a necessary (but not sufficient) condition for being.

The first context I examine is Leibniz's early distinction between infinite number and infinite being, articulated in 1675–1676. This early context is formative of Leibniz's usage of these notions as paradigms of being and non-being. The second context is Leibniz's distinction between infinite numbers, which he regards as impossible, and infinite series, which he regards as possible. This leads to the distinction between individuals and their complete concepts, which is the most important case of Leibniz's more general distinction between possible things and actual ones. Thus, I briefly examine the distinction between beings and non-beings in terms of possible things and actual ones. I then examine Leibniz's formulation of the distinction in terms of the difference between beings and beings of reason (*entia* and *entia rationis*), which will lead us to examine Leibniz's notion of aggregates as an intermediary level between beings and non-beings. After considering Leibniz's distinction between aggregates and substances (formulated and defended in the correspondence with Arnauld in 1686–1687), I finally examine his later distinction between natural machines and artificial machines (first formulated in the *Système nouveau* in 1695).

The first case is a particularly interesting case within Leibniz's broader distinction between *entia* and *entia rationis*. The final case is a particularly interesting case within Leibniz's broader distinction between aggregates and individual substances. Since these distinctions and formulations are drawn from successive stages in Leibniz's career—early, middle, and late—they illustrate a general distinction between beings and non-beings that runs in Leibniz's texts and thus exposes a general strand in the way he conceptualizes the distinction between beings and non-beings. Leibniz's distinctions between possible individuals and actual ones and his related distinction between complete concepts and created agents can only be briefly treated here.

However brief, the examination of the contexts mentioned above yields some interesting results. First, it shows that infinity, for Leibniz, is a mark of existence, that is, it is a necessary but not sufficient condition. Second, it shows that Leibniz's notion of being requires both logical and metaphysical models of substance. Recent interpretations of Leibniz's notion of substance and more generally his notion of being emphasize different aspects of his complex view. For example, in rehabilitating the tradition of the Russell/Couturaut/Gurwitsch panlogistic conception of substance, Cover and O'Leary-Hawthorne have recently argued that a Leibnizian substance is to be identified with the law of the series that underlies its complete concept (Cover and O'Leary-Hawthorne 1999, 227).[1] By contrast, Fichant (1997), Mercer (2001), and Phemister (2005) have argued that the essential aspect of Leibniz's notion of substance is its inherent capacity to act. While the logical model of substance stresses the texts in which Leibniz defines a substance through its complete concept and law of the series, the metaphysical model stresses texts in which the notions of force and activity are prominent as the defining features of substance.

My survey suggests that neither the logical nor the metaphysical model is sufficient to account for what Leibniz would consider as a being or as a complete entity. Rather, my examination suggests that each model constitutes a necessary but not sufficient condition to characterize the Leibnizian notion of being. I will suggest that both aspects are required for Leibniz's conception of a being, and that Leibniz was employing both models—logical and metaphysical—to account for the notion of being.

11.2 Infinite Number and Infinite Being

In notes and letters from 1675 to 76 Leibniz compares the notion of an infinite number and that of an infinite being. He notes that "the number of all numbers is a contradiction" (e.g., A 6.3 463; PK 7)—i.e. an impossible notion (A 6.3 477; Arthur 53).[2] Leibniz's reasoning on this point derives from Galileo's paradox, namely that the series of natural numbers cannot be equal to the series of their squares. If so, the whole (the series of natural numbers) would not be greater than their part (the series of their squares) and the fundamental axiom of the quantitative sciences would have to be given up.[3] For this reason, Leibniz concludes that the notion of an infinite number is impossible. At the same time, Leibniz famously argues that, in order to prove that the greatest or the most perfect being exists, one has to show that its notion is possible, i.e., that it does not imply a contradiction.

I have argued elsewhere (Nachtomy 2005) that Leibniz's concerns regarding the possibility of the infinite being are motivated by his concerns about the contradiction involved in the notion of infinite number. In this context, it is clear that Leibniz is investigating these notions by comparing and contrasting them, seeking to show that

[1] The abbreviations used are listed at the end of this article.

[2] In a letter to Conring (1677) Leibniz writes: "At qui subtiliores sunt adversarii ajunt Ens perfectissumum tam implicare contraditionem quam numerum maximum" (A 325).

[3] "There comes to mind a similar line of reasoning conspicuous in Galileo's writings. The number of all squares is less than the number of all numbers, since there are some numbers which are non square. On the other hand, the number of all squares is equal to the number of all numbers, which I show as follows: there is no number which does not have its own corresponding square, therefore the number of all numbers is not greater than the number of all squares; on the other hand, every square number has a number as its side: therefore, the number of squares is not greater than the number of all numbers. Therefore, the number of all numbers (square and non-square) will be neither greater than nor less than, but equal to the number of all squares: the whole will be equal to the part, which is absurd" (A 550–51; Arthur, 177). See also A 6.3. 11; Arthur 5.

maximum perfection is consistent while maximum in number is not. The contrast between them is as significant as their similarity. While both concepts ("all units", "all perfections") seem to imply infinite quantity, the concept of the infinite being serves Leibniz as a paradigm of being while the notion of the greatest number serves him as a paradigm of an impossible one.

In a Letter to Oldenburg of December 1675, Leibniz writes:

> Assuming that ... a being [whose essence is to exist] is possible or that there is some idea corresponding to these words, it certainly follows that such a being exists. But we believe that we are thinking of many things (though confusedly) which nevertheless imply a contradiction; for example, the number of all numbers. We ought strongly to suspect the concepts of infinity, of maximum and minimum, of the most perfect, and of allness (*omnitnias*) itself. Nor ought we believe in such concepts until they have been tested by that criterion which must, I believe, be credited to me, and which renders truth stable, visible and irresistible ..." (Loemker, 257)

The criterion Leibniz refers to here is his demand to provide a real definition or to give a possibility proof to such problematic notions. The case we are considering, the contrast between infinite number and infinite beings, is Leibniz's formative case for the applicability and usefulness of this criterion. Indeed, Leibniz is often using the notion of infinite number as a paradigm for an impossible notion which he *contrasts* with the possibility of the most perfect or infinite being (A 6.3 572; PK 91; A 6.3 325).[4] It need not be surprising that Leibniz is intrigued by this comparison. These concepts have a similar syntactic structure and both seem to imply an infinity or totality of simple elements: attributes or simple forms in God; simple units in number.

Moreover, Leibniz's analysis of the notion of the greatest being is carried out in terms that suggest maximal totality and number, such as "the subject of all perfections" (A 6.3 580; DSR 103), "one which contains all essence, or which has all qualities, or all affirmative attributes." In a set of definitions from 1676, (A 6.3 482–84) and in reference to Euclid's definition of number, Leibniz writes: "Number, if it is understood simply as integral and rational, is a whole consisting of units" (A 6.3; DSR 37–38). In the same group of texts he also draws an explicit analogy between God's essence and whole numbers.[5] In this analogy, numbers consist of units and God's essence consists of simple forms or perfections. If, as we have seen,

---

[4]In a letter to Conring (1677) Leibniz writes: "*At qui subtiliores sunt adversarii ajunt Ens perfectissumum tam implicare contraditionem quam numerum maximum*" (A 325). Leibniz's possibility proof is given in the following passage: "Demonstrationem reperisse videor, quod Ens perfectissimum, seu quod omnem Essentiam contineat, seu quod omnes habeat Qualitates, seu omnia attributa affirmativa, sit possibile, seu non implicet contradictionem. Hoc patebit si ostendero omnia attributa (positiva) esse inter se compatibilia. Sunt autem attributa aut resolubilia, aut irresolubilia, si resolubilia sunt erunt aggregatum eorum in quae resolvuntur; suffecerit ergo ostendisse compatibilitatem omnium primorum, sive irresolubilium attributorum, sive quae per se concipiuntur, ita enim si singula compatibilia erunt, etiam plura erunt, adeoque et composita. Tantum ergo suffecerit ostendere Ens intelligi posse, quod omnia attributa prima contineat, seu duo quaelibet attributa prima esse inter se compatibilia" (A 6.3 572; DSR, p. 91–93).

[5]See Nachtomy, (2007, chap. 1–2).

Leibniz defines whole number as consisting of units, then the greatest number would be seen as consisting of all units. Since he defines God as consisting of all essence or all perfections or all positive attributes, the greatest being is seen as consisting of all perfections. Thus, just as there are infinitely many units in the notion of an infinite number, so there would be infinitely many perfections or attributes in the notion of God.

However, if this were the case, Leibniz would have to consider these notions to be equally problematic (or equally unproblematic). Yet, he clearly doesn't. Rather, he considers the notion of an infinite being to be possible, and uses it as a paradigm of a Being, and he considers the notion of an infinite number to be impossible, and uses it as a paradigm of a non-being (not only it does not exist but it cannot exist).[6]

If the notions of all perfections and all simple units are analogous, Leibniz's position is very puzzling. Furthermore, the distinction between these notions plays an important role in other contexts in Leibniz's metaphysics. For example, we know that his later (1679–1686) notion of individual substances as well as their complete concepts involves infinity. While the complete concept of an individual involves infinitely many predicates, Leibniz's notion of individual substance involves infinitely many properties as well as relations to infinitely many other individuals. Since he considers individual beings to be possible (as some of them are actual), surely he considers their infinite concepts to be non-contradictory. But, as just noted, he considers infinite number to be impossible. So, to put the question more generally, what is the difference Leibniz sees between the notion of an infinite being (or substance)—created or not—and that of an infinite number? How can he justify his claim that the notion of infinite being is non-contradictory if that of infinite number is contradictory?

One might suggest that an important difference Leibniz sees between these notions derives from the semantic difference, namely that the one is a notion of a being and the other is a notion of a non-being. While numbers are universal, divisible, and may be composed by a conjunction of units; beings, for Leibniz, are individual (that is, unique), active and indivisible. While beings for Leibniz must be active agents, numbers are not active. Rather, numbers are understood as abstractions in the minds of agents.

Unlike the notion of a number, the notion of God (and, in fact, of any true being, of which God is the most perfect) is, according to Leibniz, a notion of something that is not produced by composition of parts, that is, it is not something that is made up (per impossible) by composing or conjoining an infinite number of units or perfections. God (and any created Leibnizian being) is not a sum of perfections. In fact, a being for Leibniz is not a sum at all; rather, it is an active agent and, in this sense, it is one and indivisible. The whole agent acts and his action is not a sum of

[6]In this period (as well as later ones, cf. 1984; *Discours de métaphysique*, New Essays), it is clear that Leibniz is investigating these notions by comparing and contrasting them. It is arguable that Leibniz's concern regarding the possibility of the infinite being (and perhaps of possibility proofs in general) is driven by his concerns about the contradiction he discerns in the notion of infinite number, most rapid motion, and its likes. See my (2005).

the actions of its constituents. Such a unity cannot be fully defined in terms of its constituents and in this sense admits of no parts. If so, one clear difference between numbers and beings is that a being has unity and activity while a number does not. In other words, a number is not an entity because it is neither one nor active. And so, even if it is unclear whether Leibniz is entitled to this distinction (see Nachtomy 2005), it seems quite clear that, for him, not only infinity but also activity and unity must characterize being. As we shall see, this conclusion is supported by the other contexts of our survey.

Another interesting question that arises from the comparison between the notion of infinite number and that of infinite being is whether the kind of infinity Leibniz ascribes to complete beings and true units is the same as that he ascribes to non-beings. My conjecture is that it is not. I will examine and reformulate this conjecture in the course of this paper through a number of examples and contexts I now turn to consider.

11.3 Infinite Number and Infinite Series

Let me begin with a particularly illuminating example. While Leibniz rejects the notion of an infinite number as contradictory, he accepts the notion of infinite series of numbers. Indeed, the notion of a series plays a central role in his development of the infinitesimal calculus. This gives rise to the following question: What is the difference Leibniz sees between infinite series and infinite series of numbers, such that the one is consistent and acceptable while the other is contradictory?

One crucial difference can be clearly stated: the notion of an infinite series avoids the contradiction of an infinite number because it is not defined as a collection of units but rather according to its law of generation. A series is not defined as a sum of units but according to its rule of production—a kind of machine whose activity produces well-regulated results.[7] Leibniz employs here his notion of generative definition, which he is also using to provide real definitions, that is, in demonstrating that a given concept is consistent. In this sense, to define x is to construct a concept of x, which also demonstrates its self-consistency (just as we have seen that Leibniz demands in the context of proving the possibility of the *Ens Perfectissimum* and disproving that of an infinite number).

Let us examine the difference in Leibniz's approach to infinite numbers and infinite series in a bit more detail. For simplicity's sake, we can contrast two very intuitive notions, the notion of an infinite number and the series of natural numbers. It is crucial to observe that Leibniz does not see the series of natural numbers as a collection of units but rather gives it an operative definition. His definition is given through a procedure that *generates* the series, so that the successive number results from the addition of "one" (cf. e.g., New Essays 4.2.1). The reiterability of this procedure implies that it can be carried on without limitation. For Leibniz, this

[7]"A series is a multitude with a rule of order" (A 6.4 1426).

shows the intelligibility of the series of natural numbers. But it also reveals something essential to Leibniz's analysis of infinity in this context, namely that it is bound up with a notion of activity or, more precisely, with a possible activity of an agent, namely, the possibility of continuing to apply the procedure. Thus, while an infinite number is not seen as a sum of units, i.e., it is not a whole, one can speak of the law generating a series, seen as a definition *in act*, and thus as what gives the series its unity and intelligibility. Thanks to its formation law, one can speak of a series rather than of an aggregate of units.

This point reveals a profound aspect about the way Leibniz conceives of numerical infinity and the extent (as well as the exact sense) of its intelligibility. It is instructive to recall Yvon Belaval's insightful remark on this point. Belaval has suggested that Leibniz's definition of "number" is operative. According to Belaval, this definition implies that

> le nombre, tant que relation, n'a de réalité (et de possibilité) que pour autant qu'il est pensé. C'est donc du côté de l'esprit qu'il faut chercher la source de l'infinité numérique: d'une manière plus précise, c'est l'infinité actuelle de l'esprit qui rend compte de l'infinité virtuelle de toute suite numérique. (Belval 1960, 270)

Belaval's remark goes down to the heart of Leibniz's views on infinity and being. The way Belaval presents this point shows that numerical infinity can only be placed in the context of the actual thoughts of an active mind. In this sense, the reality of infinite (in the quantitative sense) notions presupposes the reality of the mind thinking them. This sort of conceptualism about abstract entities such as number and relations, is a familiar theme in Leibniz.[8] Yet the far-reaching consequences—especially for our current purposes of distinguishing between infinite beings and infinite non-beings—are not sufficiently appreciated. In light of this definition, the series of natural numbers can be seen as infinite to the extent that it presupposes a mind that would go on to add units. In other words, the method of production presupposes a producing (thinking) agent.

This conceptualist approach explains why Leibniz argues that infinity is to be regarded as merely potential in abstract and mathematical contexts. Such a potential type of infinity does not apply to the infinity of beings, which are active thinking agents, whose activity is non-discrete and whose infinity (and perfection) is not potential but actual. This difference corresponds to the qualitative feature of active agents as distinct from the quantitative feature of a number as a sum of units. Hence, in the realm of beings, infinity is regarded not *in potentia* but rather *in act*. By contrast, in the realm of non-beings, infinity is understood as potential.

---

[8]As Couturat nicely points out (1961, 471; my translation), "one can say that Leibniz remains a nominalist in an entirely negative sense, namely that he rejects realism and denies universals a real and substantial existence. But he does not thereby refuse to assign them objective value, like the nominalists who reduce them to names. Rather, he adapts and intermediate position, which one designate by the name *conceptualism*, ... " As Mugnai notes (1992, 25), "there are no ideas without the intellectual activity of someone thinking (be it God or man or some other rational being)."

This point gives rise to a neat division: the type of infinity Leibniz ascribes to beings is different from that he ascribes to non-beings. In particular, one type of infinity applies to thinking beings and another applies to thoughts. While the one is quantitative and discrete, the other is qualitative and indivisible. This difference is strongly related to a metaphysical distinction. We can now better describe the origin of this division and its place in Leibniz's metaphysics, namely as rooted in the difference between actual thinkers and potential thoughts. As we shall see, this point reoccurs in a variety of other context in Leibniz's metaphysics.

11.4 Complete Concepts of Individuals and Created Individuals

Like the generative definition of the natural numbers, Leibniz also uses the notion of the "law of a series" at one of the most central points of his metaphysics. The notion of the law of a series is used to define not only series of numbers but also beings—that is, individual substances. As is well known, Leibniz holds that every created being, that is, every individual substance, is defined through its own law of the series. This indicates at once that all Leibnizian beings other than God are also infinite and are defined and individuated through their unique law of generation.[9]

At the same time, we also know that, according to Leibniz, each individual has a concept so complete that it entails all its predicates. In other words, the complete concept of the individual includes anything true of it (*Discours de métaphysique* 8 and 13). It goes without saying that such a concept is infinite. This gives us another sharp contrast between an infinite being (an individual) and an infinite non-being or its conceptual counterpart. In other words, this is yet another clear articulation in Leibniz's metaphysics of the contrast between a concept and an agent. It goes without saying, too, that a concept of an individual must be consistent. This is obvious because, if an individual exists, as many surely do, then they must also be possible, that is, they must have consistent concepts or perfect representations in God's understanding. And, indeed, this is what Leibniz illustrates with his famous examples of the concepts of Alexander and Caesar in the *Discours de métaphysique* and the ensuing correspondence with Arnauld.

Leibniz, however, is somewhat less explicit about how such infinite concepts of individuals are to be defined. Since I have written about this topic at length elsewhere, let me just emphasize one point here. Leibniz is very explicit that the concept of an individual involves (or includes) infinitely many predicates, which all serve to define the whole career of an individual—past, present, and future. Thus, each predicate is an essential component of the definition of that individual. However, if such a concept would be seen as a mere conjunction of predicates or as a set of infinitely

[9]"The law of order ... constitutes the individuality of each particular substance" (GP IV, 518; L 493). "For me nothing is permanent in things except the law itself ... The fact that a certain law persists, which involves the future states of what we conceive to be the same – this is the very fact, I say, that constitutes that same substance" (GP II 263–64; L 534–35). See also *Theodicy* 291.

many predicates it would turn out to be contradictory, just as the notion of an infinite number of units. For this reason, it seems likely that Leibniz thought about the infinite concept of an individual along the lines that he thought about an infinite series, i.e., as defined by its law of generation.[10]

Indeed, I would suggest that this is why Leibniz has found the notion of the law of the series attractive in the first place. I shall argue later that the law of the series suffices to define the concept of an individual but that it is not sufficient to define an actual individual. An actual individual also involves inherent force as a source of the individual's activity. (More on this later.)

In fact, this comparison between the complete *concept* of an individual and an actual individual illustrates the most fundamental case in Leibniz's metaphysics, namely a case in which both beings and non-beings are infinite and both play an indispensable role in defining the things that exist in our world. One might say that this comparison brings us to the very heart of the positive aspect of Leibniz's metaphysics. At the same time, it shows the double face of Leibnizian beings: on the one hand, they are defined through consistent individual concepts; on the other, they only exist as active agents.

11.5 Possible Things and Actual Things

This point is strengthened when we note that the contrast between individuals and their complete concepts expresses a more general distinction in Leibniz's metaphysics, namely that between possible things and actual ones. There is no need to stress here the centrality of these notions in Leibniz's metaphysics. Let me just remark that, after all, complete concepts are used as a way to talk about and make sense of the notion of possible individuals, which, if actualized, become individual substances. As we know, for Leibniz such substances make up the ontology of the created world. Whatever else there is in the created world (including extension, space, and time) is parasitic on the existence of individual substances (and, for this very reason, extension, time, and space are not considered as true beings).

At the same time, it is worth noting that the space of possibilities, for Leibniz, is broader than that of concepts of individuals. For example, it includes concepts of individuals that are merely possible (i.e., which will never be actualized); possible worlds which will never be created; concepts, relations, and proportions, which are merely intelligible (that is, consistent) but, since they are not concepts of individuals, they are not even candidates for actualization. Since the realm of the possible includes any idea or thought that is non-contradictory, it is obvious that the space of possible things is far more inclusive than that of actual ones. In particular, it obviously includes both infinite and finite notions. This illustrates that we find both infinite and finite items in the realm of non-beings or possibilities, which are

[10]See Nachtomy (2007, chaps. 1–2).

merely conceived in the understanding—either human or divine. By contrast, true beings, for Leibniz, are always infinite. This brings us to another fundamental way to articulate the distinction between being and non-beings in Leibniz's metaphysics.

11.6 Entia and entia rationis

The traditional distinction between *entia* (beings) and *entia rationis* (beings of reason) is very instructive for clarifying the way Leibniz articulates the distinction between beings and non-beings, and especially so in clarifying which kind of infinity is applicable to each one of them. While this distinction is a commonplace, it is very interesting to observe the role it plays in the context of Leibniz's metaphysics.

It is first important to observe that the realm of *entia rationis* for Leibniz is not only immense but also includes the objects of some of the most important sciences. Thus, possibilities, concepts, and relations are classified as *entia rationis,* but also the objects of mathematics (numbers, relations, proportions) and geometry.[11] It is clear that *Entia rationis,* for Leibniz, are not fictions or mere figments of the imagination. As we have seen, they are not impossible entities but rather consistent thoughts and concepts (either human or divine). Thus numbers, shapes, and, in effect, all universal concepts are for Leibniz incomplete beings. Such incomplete beings are either abstractions from concrete beings or constructions of simple beings.

Thus, it turns out that for Leibniz, some non beings, such as series, lines, relations, proportions are mental abstractions that may develop (or divide) to infinity. This is so because they are considered in the first place as products of some thinking beings, hence, as abstractions and thoughts rather than as concrete things. Thus a series such as ½, ¼, 1/8, 1/16 ... is infinite in the sense that someone could go on, according to a rule. But beings, such as God and created substances are infinite in a different sense, namely, they are infinite in act, not in *potentia* because they are concrete indivisible and perfect wholes, either in their kind (in the case of created substances) or absolutely perfect (in the case of God).

This nicely explains Leibniz's articulation of the distinction between ideal and real, abstract and concrete things, such that, while possible things admit of potential infinity, existing things admit of actual infinity. This distinction is usually drawn in the context of divisibility to infinity. But it seems to apply more generally, so that it cuts through infinite beings and non-beings. Abstract and ideal things are potentially infinite; concrete and real things are actually infinite. As I noted, however, they are not infinite in the same sense: while abstract and ideal things are infinite in a quantitative (and divisible or discrete) sense; concrete and real things are actually infinite in the qualitative (indivisible and non discrete) sense of activity and perfection.

---

[11] "Numbers, modes, and relations are not entities" (A 6.3 463; DSR 7).

Along these lines, it seems that we can distinguish between

1. Infinity of thoughts, which is abstract and *in potentia*, and pertains to concepts, series, ideas, relations, possibilities, and other *entia rationis*.
2. Infinity of thinking (beings), which is concrete and *in act*, and pertains to substances alone, either created or not.
3. An intermediate level: infinity of thoughts and thinking beings, which is semi mental and semi concrete. This level pertains to aggregates and relations, which are founded in a multiplicity of real things but whose unity derives from a mind perceiving them all together, which is why they are considered to be semi mental and semi real. The final two sections of my survey in this article are dedicated this intermediate level.[12]

11.7 Aggregates and Substances

As I noted, the unity of beings and their non-compositional nature play an essential role in Leibniz's distinction between created, individual substances, which are one and active, and aggregates, which are neither one nor act as one agent. According to Leibniz, aggregates, such as woodpiles and armies, are beings by aggregation and therefore are not truly one thing. While one may ascribe activity to their constituents, they do not act as a single agent but as an aggregate of many agents. In fact, it seems more precise to say that they do not act as a single agent. By contrast, true beings are not composed and are truly one. For example, Leibniz writes that "... no entity that is truly one [*ens vere unum*] is composed of parts. Every substance is indivisible and whatever has parts is not an entity but only a phenomenon."[13]

Let us now attend to the difference between the way a number is said to result from a composition of units (which we have considered in the first section) and the way an aggregate results from a composition of parts. A major difference stems from the fact that what is being composed (or aggregated) in aggregates are real constituents whereas, in numbers (as well as other abstractions), what is being composed are possible (but not actual) constituents or parts. This distinction becomes all the more significant in the context of infinity. Unlike a number (and an aggregate), a being, for Leibniz, is not defined quantitatively or compositionally; rather, it is defined through a basic capacity to act or what Leibniz calls its primitive force (or to which he sometimes refers to as its Entelechy).

Yet, as Levey notes, there are also close connections between Leibniz's approach to the question of infinite number and that of aggregates.

---

[12]There are two other important distinctions that I can only mention here: (1) Leibniz's distinction between complete and incomplete beings, and (2) his distinction between the abstract and the concrete.

[13]Quoted in Brown (2000, 41).

Leibniz construes cardinal numbers (other than 1 and 0) as aggregates of 'unities' or 'ones' (e.g., $6 = 1 + 1 + 1 + 1 + 1 + 1$) and he views them as applying to aggregates of things taken as a whole rather than to uncollected things taken as so many individuals. Where the number 1 or unity applies to an individual thing, the number 5 applies to an aggregate 'whole' with five 'parts' (what we would probably consider to be a *set* with five *members*). So in his view, the infinite cardinal number is bound up with the concept of an infinite aggregate or whole in two ways. First, an infinite number itself would count as an infinite whole; and second, such a number would apply to infinite aggregates or wholes. Consequently, the concept of an infinite number is also doubly bound up with Galileo paradox: both the number and what it *numbers* would violate the part-whole axiom, if either were to exist." (Levey 1998, 103)

Leibniz denies that any of these exists precisely because neither can be considered to be a whole. As he writes, "an infinity of things is not one whole, or [. . .] there is no aggregate of them" (A 6.3 503), which is in contrast with a passage referring to God, namely, that, "an infinite whole is one" (A 6.3 474; Arthur 49), and in contrast with Leibniz's claim that an organic being consists of infinitely many others. While Leibniz does not deny that we can talk about infinitely many things (if only syncategoretmatically), he denies that an aggregate of such things is a being.[14] In this respect, the distinction between an infinite number and an infinite being seems to be analogous to the distinction between aggregates, which Leibniz does not regard as true beings, and individual substances, which he regards as true beings. This is the case because beings are not aggregated or composed but are indivisible active units. This is explained through the definition of individuals by their inherent laws of action which are one and indestructible. They are not composed or constructed of more basic constituents but are created as active infinite beings whose constituents can be considered by abstraction (but not by decomposition).[15]

The contrast between beings and non-beings gives rise to Leibniz's three-fold distinction between indivisible individuals, real or well-founded phenomena, and mere phenomena, which are incomplete *entia rationis*.

A *suppositum* is either an *individual* substance, which is a complete entity, one in itself, such as God, a mind, the ego; or it is a real phenomenon, such as a body, the world, a rainbow, a woodpile. We conceive the latter on the model of a complete substance, but since body—unless it is animated, or contains within it a certain single substance, corresponding to the soul, which they call a substantial form or primary entelechy—is no more one substance than a woodpile; and since again there is no part of it which can be regarded as a unity in itself (since body is actually subdivided, or certainly subdivisible, into parts), it is a consequence that every body will only be a real phenomenon, like a rainbow. Similarly mathematical things, such as space, time, a sphere, an hour, are merely phenomena, which we conceive on the model of substances. And accordingly there is no real substance which is not indivisible one. And indeed, it may be that those things that are divisible and consist in magnitude, such as space, time, and bulk, are not complete beings, but must have something

[14]In his mature philosophy, Leibniz defines the world as an aggregate of finite things (*aggregatum rerum finitarum*) (cf. GP VII, 302). Similarly, in his letter to Gabriel Wagner of March 3 1698, Leibniz defines the world as an aggregate of changeable things or things that are susceptible of imperfection (cf. Grua, 396; Adams 1994, 15).

[15]I believe that this is the reason why what Levey has recently called "the construction problem" rests on a misconstrued version of Leibniz's view of being and unity (Levey 2007, 64–66).

superadded to them, which involves all those things that can be attributed to this space, this time, this bulk. (A 6.3 132; Arthur 265–67)[16]

As we have seen above, Leibniz does not deny that infinity applies to complete beings. Rather, according to Leibniz, infinity is one of the defining features of beings—both created and not. For example, we know that he characterizes organic beings as nesting infinitely many such beings, and we know, furthermore, that his very distinction between artificial machines, which he does not consider as true beings, and natural machines, which he considers as true beings, turns on the fact that natural machines have a nested structure to infinity (see next section). But, as we have also seen, the sense in which infinity qualifies beings cannot apply to the quantitative aspects of magnitude and the number of their constituents (for otherwise they would be as contradictory as the greatest number). It is partly for this reason that, as we have seen, Leibniz does not define created beings as sums of their predicates or as combinations of their constituents but rather through their unique method of production or their law of formation.[17] Since Leibniz is fully aware of the impossibility of an infinite sum of predicates or perfections, he does not identify the principle of individuation with a unique sum of predicates but with its method of production. Leibniz's genetic definition of beings explains how a thing is produced through its formation law.[18] By defining a being through its formation law Leibniz can hold that a being is both one and infinite at the same time.

However, as we have seen, Leibniz applies genetic definitions to mathematical things such as infinite series as well as to concepts of individuals. As Couturat already pointed out, Leibniz defines infinite series according to their laws of production. By Leibniz's lights, an infinite series is, like a number, a mathematical abstraction, a being of reason, not a true being. Hence, a law of production is not sufficient to distinguish between infinite beings, such as Leibnizian individuals, and infinite non-beings or abstractions, such as mathematical series, possible individuals or concepts of individuals. This observation is supported by Leibniz's practice to define concepts of individuals (which are clearly distinguished from existing individuals) by their laws of production.[19] Since the law of production is applicable to concepts as well, it must be regarded as a necessary but not sufficient condition for

[16]"The aggregate of all bodies is called the *world*, which, if it is infinite, is not even one entity, any more than an infinite straight line or the greatest number are. So God cannot be understood as the *World Soul:* not the soul of a finite world because God himself is infinite, and not of an infinite world because an infinite body cannot be understood as one entity, but that which is not an entity in itself has no substantial form, and therefore no soul. So Martianus Capella is right to call God an extramundane intelligence" (A 6.4 1509; Arthur, 287).

[17]A corollary to this view is Leibniz's definition of infinite series. He does not define infinite series as a sum of numbers but as a product of its formation rule. In this connection, see Couturat's interesting discussion (1973, 476). Couturat cites this passage from the letter to des Bosses (of 11 March, 1706): "Neque enim negari potest, omnium numerorum possibilium naturas revera dari, saltem in divina mente, adeoque numerorum multidudinem esse infinitiam."

[18]See Gurwitsch (1974, 65–72), section II d., on Generative Definitions.

[19]I have argued for this point in my (Nachtomy, 2002, 31–58).

characterizing Leibnizian beings. In addition, according to Leibniz, a being must also be active and entail its source of activity. Nonbeings, such as infinite series and possibilities, are not active but are rather conceived by a different mind—divine or human. Hence, they are regarded as *potentially* infinite. They *can* be divided or be composed to infinity. This is consistent with Leibniz's acceptance of infinite beings that are true substances and his conception of infinite numbers as impossible beings, numbers as *entia rationis*, and aggregates as semi-beings or well founded phenomena. Much more needs to be said about these distinctions than I have space here. But let me just note that the notion of syncategormatic infinite is available for Leibniz to characterize such concrete multiplicities as aggregates.[20]

Leibniz's acceptance of infinite beings (God and individual substances) and his rejection of an infinite number (and the above observations) seem to indicate that activity, unity and infinity are all necessary requisites for being, according to Leibniz. In addition, his rejection of aggregates as true beings indicates that infinity is not the distinguishing feature between aggregates and substances since both involve an infinity of substances (even if, as hinted above, the infinity ascribed to substances and aggregates is of different sense). What accounts for the distinction between aggregates and substances seems to be their source and type of unity. While the unity of aggregates is external, the unity of individual substances, which also consist of infinitely many constituents is intrinsic. This point is made even more explicit in the context of Leibniz's distinction between natural/organic machines and artificial ones to which I now turn.

11.8 Natural Machines and Artificial Machines

As Fichant notes, Leibniz introduces the concept of "natural machine" in the *Système Nouveau de la nature* (1695) as a mean to limit the claims of a mechanistic approach, which confuses the natural with the artificial. Particularly relevant is Descartes's program to describe animals as subtle machines that lack any internal power.[21] By contrast, Leibniz's agenda may be seen as an attempt to revive the Aristotelian distinction between animate and inanimate things in "an intelligible

[20] For the origin and the meaning of this doctrine, see Richard Arthur's introduction to *Arthur*.

[21] In his *Principles of Philosophy* article 203, Descartes seems to assimilate the artificial and the natural. For him, artificial machines serve as models to explain the natural ones. Natural machines are like artificial ones, except much more complicated. He wants to establish that they are of the same kind. He uses the notion of divine created machines to show that the subtle parts of machines are extremely complex and invisible to us. While both Descartes and Leibniz argue that machines are extremely subtle, Descartes uses this point to argue for his view that, in the final analysis, animals are nothing but subtle machines. By contrast, Leibniz uses this point to argue that there is a categorical difference between them. See also *Les passions de l'ame,* first part, articles 5 and 6 where he writes e.g., that the body has in it "the corporeal source of movement" (art. 6).

way" and resist the reduction of natural machines to artificial ones.[22] It is with this aim in mind that Leibniz draws the distinction between artificial and natural machines in the System Nouveau. Fichant points out that, according to Leibniz, the difference between nature and art is marked by two traits: composition to infinity, which guarantees indestructibility, and true unity, which is the foundation of substantiality in natural machines (Fichant 2003, introduction). As Leibniz writes,

> Il faut donc savoir que les machines de la nature ont un nombre d'organes infini, et sont si bien munies et à l'épreuve de tous les accidents, qu'il n'est pas possible de les détruire. Une machine naturelle demeure encore machine dans ses moindres parties, et qui plus est, elle demeure toujours cette même machine qu'elle a été, n'étant que transformée par des différents plis qu'elle reçoit, et tantôt étendue, tantôt resserrée et comme concentrée lorsqu'on croit qu'elle est perdue.
>
> De plus, par le moyen de l'âme ou forme, il y a une véritable unité qui répond à ce qu'on appelle MOI en nous; ce qui ne saurait avoir lieu ni dans les machines de l'art, ni dans la simple masse de la matière, quelque organisée qu'elle puisse être, qu'on ne peut considérer que comme une armée ou un troupeau, ou comme un étang plein de poissons, ou comme une montre composée de ressorts et de roues. (GP IV, 482)

I have suggested above that these two traits—infinity and unity—derive from a common source—viz., the formation law of the individual, which unifies its infinite constituents. As this text indicates, the distinction between natural and artificial machines is a particular case within the general distinction between aggregates and substances. But while Leibniz wishes to stress the difference between the mechanic and the organic, he is very explicit that the only difference between natural machines and artificial ones is the nested structure to infinity of the natural ones. This difference is also related to the intrinsic unity and substantiality of natural things (see GP III, 457). In accordance with Leibniz's commitment to an intrinsic connection between being and unity, the structured ensemble of natural machines must have substantial unity or else it would not differ from mere aggregates.[23] This implies that a natural machine must be united as *one* single substance, as Leibniz states in a letter to De Volder:

> Although I said that a substance, even though corporeal, contains an infinity of machines, at the same time, I think that we must add that a substance constitutes one machine composed of them, and furthermore, that it is activated by *one* entelechy, without which there would be no principle of true unity in it. (AG 175; my italics).[24]

While the particular distinction between artificial and natural machines seems to turn on a nuance, it is in effect highly consequential. One obvious consequence is

---

[22]See for example, Leibniz's controversy with Stahl (Carvallo 2004, 80), where Leibniz criticizes the Moderns for pretending that "*nihil aliud sit natura corporum quam Mechanismus*" (there is nothing in the nature of bodies but mechanism).

[23]". . . since I am truly a single indivisible substance, unresolvable into any others, the permanent and constant subject of my actions and passions, it is necessary that there be a persisting individual substance over and above the organic body" (Comments on Fardella, AG 104).

[24]See also GP II 252; GP VII 502 and C 13–14.

that a machine of nature is identified with a corporeal substance in counter distinction from an aggregate.[25] Indeed, most of the differences between aggregates and individual substances apply to the distinction between artificial machines and natural ones, which indicates that Leibniz regards the two distinctive traits of natural machines—infinity and unity—as distinctive features of beings. But this is not the place to expand on this fascinating topic.[26] It is rather time to conclude this paper.

11.9 Conclusion

The distinctions and contrasts considered above support the following general conclusion: a being, according to Leibniz, requires both a law generating an infinite series and primitive force or power of action. The first set of examples, contrasting an infinite number with an infinite being, shows that, while both notions are infinite, an infinite being does not have a compositional or additive unity but rather requires primitive unity. While this is to be expected in an example that contrasts real beings with ideal numbers, we see similar reasoning (and conclusion) in the context of the real and concrete things, that is, in the contrast Leibniz draws between aggregates and individual substances. The unity of individual substances is law like and internal, the unity of aggregates is external in requiring a mind to perceive the various relata (e.g., sheep) as one thing (e.g. flock). This point is confirmed by the fact that Leibniz's distinction between artificial and natural machines turns on the infinite nested structure of the natural, whose law of generation also functions as its source of unity. As the first contrast makes clear, as well as Leibniz's view of actualization and many texts confirm, a law is a necessary (but not sufficient) condition for a complete Leibnizian being—power of action or primitive force are required as well. In this sense, both models—logical and metaphysical—are required for Leibniz's conception of being. While the logical model may account for the notions of some infinite non-beings, such as infinite series and concepts of individuals, power and activity are indispensable aspects of a Leibnizian being as well.

Abbreviations

11.9.1 References to Leibniz Works

A *Sämtliche Schriften und Briefe,* ed. Deutsche Akademie der Wissenschaften zu Berlin (Darmstadt, 1923 ff., Leibzig, 1938 ff., Berlin, 1950 ff). Cited by Series (Reihe) and Volume (Band). References in text: (A 6.3 456).

[25] Je ne compte pour substances corporelles que les machines de la nature qui ont des âmes ou quelque chose d'analogique; autrement il n'y aura point de vraie unité (A Jaquelot, 22 mars 1703, GP III, 457).

[26] See Fichant (2003) and Duchesneau (1998).

AG *Philosophical Essays*, trans. and ed. Roger Ariew and Daniel Garber (Indianapolis, 1989).

Arthur G. W. Leibniz, *The Labyrinth of the Continuum. Writings on the Continuum Problem, 1672–1686*, edited and translated by Richard Arthur, (New Haven and London: Yale University Press, 2001).

C *Opuscules et fragments inédits de Leibniz. Extraits des manuscrits*. ed. Louis Couturat (Paris, 1903, repr. of Hildesheim, 1961).

Carvallo *La controverse entre Stahl et Leibniz sur la vie, l'organisme et le mixte: doutes concernant la vraie théorie* médicale *du célèbre Stahl, avec les répliques de Leibniz aux observations stahliennes*, ed. Sarah Carvallo. Paris: Vrin, 2004.

DSR *De Summa Rerum: Metaphysical Paper, 1675–1676*, trans. and ed. G.H.R. Parkinson (New Haven, *1992*).

GP *Die philosophischen Schriften*, 7 vols, ed. C. I. Gerhardt (Berlin, 1875–90, repr. of Hildesheim, *1965*).

Grua *Textes inédits d'après les manuscrits de la Bibliothèque provinciale de Hanovre*, 2 vols, ed. Gaston Grua (Paris, 1948).

L *Philosophical Papers and Letters*, ed. and trans. Leroy E. Loemker, 2nd ed., (Dordrect, 1969).

References to Secondary Sources

Belaval, Yvon. 1960. *Leibniz critique de Descartes*. Paris: Gallimard.

Brown, Gregory. (2000, 41). Leibniz on wholes, unities, and infinite number. *The Leibniz Review* 10: 21–51.

Cover, J. A. and J. O'Leary-Hawthorne. 1999. *Substance and individuation in Leibniz*. Cambridge: Cambridge University Press.

Couturat, Louis. 1961. *La logique de Leibniz*. Hildesheim: Olms.

Couturat, Louis. 1973. *De l'infini mathematique*. Paris: Librairie Scientifique et Technique Albert Blanchard.

Duchesneau François. *Les modèles du vivant de Descartes à Leibniz.*, Paris: Vrin, 1998.

Fichant, Michel. 1997. L'ontologie leibnizienne de l'action: Actiones sunt suppositorum. *Philosophie* 53, pp. 135–48.

Fichant, Michel. 2003. Leibniz et les machines de la nature. *Studia Leibnitiana* 35: 1–28.

Gurwitsch, Aron. 1974. *Leibniz. Philosophie des Panlogismus*. New York: De Gruyter.

Levey, Samuel. 1998. Leibniz on mathematics and the actually infinite division of matter. *Philosophical Review* 107: 49–96.

Levey, Samuel. 2007. On unity and simple substance in Leibniz. *The Leibniz Review* 17: 61–106.

Mercer, Christia. 2001. *Leibniz's metaphysics: Its origin and development*. Cambridge: Cambridge University Press

Mugnai, Massimo. 1992. *Leibniz's theory of relations. Studia Leibnitiana Supplement 28*. Stuttgart: F. Steiner.

Nachtomy, Ohad. 2002. Leibniz on possible individuals. *Studia Leibnitiana* Band 34(1): 31–58.

Nachtomy, Ohad. 2005. Leibniz on the greatest number and the greatest being. *The Leibniz Review* 15: 49–66.

Nachtomy, Ohad. 2007. *Possibility, agency, and individuality in Leibniz's metaphysics*. Dordrecht: Springer.

Phemister, Pauline. 2005. *Leibniz and the natural world. Activity, passivity and corporeal substances in Leibniz's philosophy*. Dordrecht: Springer.

Chapter 12
Grounding the Principle of Sufficient Reason: Leibnizian Rationalism and the Humean Challenge

Brandon C. Look

12.1 Introduction

If anything counts as a fundamental tenet of "rationalism", it is surely the *Principle of Sufficient Reason* (PSR). We know it in various guises: *nihil est sine ratione; ex nihilo nihil fit;* there is nothing without an explanation; nothing comes from nothing; or every event has a cause.[1] And we can find it in all the great rationalist thinkers. It plays a prominent role in Descartes's argument for the existence of God in *Meditation III* in the form "something cannot arise from nothing"[2] and is presented as an innate idea in the *First Set of Replies*: "the light of nature does establish that if anything exists we may always ask why it exists; that is, we may inquire into its efficient cause, or, if it does not have one, we may demand why it does not need one."[3] Spinoza likewise employs the Principle of Sufficient Reason in his proof for the existence of God: "For each thing there must be assigned a cause, *or* reason, as much for its existence as for its nonexistence."[4] And, finally, in Malebranche: "there cannot be an effect or a change without a cause."[5] In short, the world is ordered and intelligible, and there is an explanation for everything. When the Principle of Sufficient Reason is taken to entail the Universal Causal Principle, then, it seems it can be found in many of the canonical western thinkers extending all the way back to Parmenides. But it is in Leibniz's philosophy, of course, that the Principle of Sufficient Reason features so prominently and in Humean skepticism that it is thought to meet its downfall.

B.C. Look (✉)
University of Kentucky, Lexington, KY, USA
e-mail: look@uky.edu

[1]One should be careful about equating the Principle of Sufficient Reason with a Universal Causal Principle. See Carraud (2002) and Pruss (2006). But, for my purposes here, I shall treat the CP—every event has a cause—as a species of PSR—everything has a reason (why it is).

[2]AT VII 40/CSM II 28. AT = Descartes (1964–74) and CSM = Descartes (1984).

[3]AT VII 108/CSM II 78. The extraordinary case of a thing not having an efficient cause outside of itself is, of course, God.

[4]*Ethics*, Part I, Proposition 11, Definition 2, from Spinoza 1985.

[5]Malebranche (1958–68), Volume XII, p. 175.

C. Fraenkel et al. (eds.), *The Rationalists: Between Tradition and Innovation*, The New Synthese Historical Library 65, DOI 10.1007/978-90-481-9385-1_12,

Considering the matter more carefully, we should say that Leibniz made explicit a principle at work in the whole tradition of philosophy and scientific thought. After all, Leibniz famously writes in the *Monadology* that all our reasonings are based on two great principles: the Principle of Contradiction (PC) and the Principle of Sufficient Reason. And he seems to treat these principles as equally fundamental, the one grounding necessary truths, the other grounding contingent truths. But one research project that seems to have been largely forgotten in the history of philosophy is the attempt, largely on the part of Leibniz's eighteenth-century German rationalist followers, to *derive* PSR from PC. In other words, PSR was taken to be a logical consequence of PC. And while Leibniz, in the majority of his writings, seems to treat both principles as equally self-evident, even he does try to independently justify PSR and to derive PSR from PC and his concept containment account of truth, or so I shall argue. If such arguments are successful, they might be able to undermine challenges to the Principle of Sufficient Reason or the Universal Causal Principle, for they will show that the legitimacy of PSR is not simply a matter of competing intuitions and claims to self-evidence. On the other hand, Hume himself tried to undermine arguments for PSR and CP in the *Treatise of Human Nature*, through arguments that stem mainly from his empiricist predecessors, Hobbes, Locke, and Samuel Clarke. Indeed, according to Hume, "every demonstration which has been produc'd for the necessity of a cause, is fallacious and sophistical." (*THN* 1.3.3.4)[6].

In this paper, then, I shall analyze two arguments that Leibniz gives purporting to justify or establish the truth of the Principle of Sufficient Reason: the first, from one of his earliest pieces; the second from his work in the 1680s, principally from the *Primary Truths*. I shall next turn to arguments given by his rationalist successors Wolff and Baumgarten—arguments that have generally been considered abject failures. Next, I shall examine Hume's challenge to the Universal Causal Principle and argue that Hume's argument does not rule out all of the rationalist arguments for the Principle of Sufficient Reason. Ultimately, I would like to show that the acceptance of "brute facts" (other than in perhaps matters of quantum physics) is due mainly to a *faux metaphysical machismo* that even Hume rejected. In other words, I plan to argue that there is indeed still something to be said in favor of the Principle of Sufficient Reason.

[6]In this paper, the following abbreviations will be used: "A" followed by series, volume, and page number = Leibniz (1923–) ; "AG" = Leibniz (1989); "C" = Leibniz (1903); "CP" = Leibniz (2005); "DSR" = Leibniz (1992); "G" followed by volume and page number = Leibniz (1875–90); "H" = Leibniz (1985); "L" = Leibniz (1969); "LLP" = Leibniz (1960); "RB" = Leibniz (1981). References to Hume's *Treatise* are to Hume (1739–40/2007); I shall simply abbreviate this "THN" followed by book, part, section, and paragraph numbers. Finally, "AA" = Kant (1902–1983), followed by volume and page number.

12.2 Leibniz

It is well-known that, according to Leibniz, the Principle of Sufficient Reason plays a fundamental role in all philosophizing. As he famously puts it in §§31–32 of the *Monadology*,

> Our reasonings are based on *two great principles, that of contradiction,* in virtue of which we judge that which involves a contradiction to be false, and that which is opposed or contradictory to the false to be true. And *that of sufficient reason,* by virtue of which we consider that we can find no true or existent fact, no true assertion, without there being a sufficient reason why it is thus and not otherwise, although most of the time these reasons cannot be known to us. (G VI 612/AG 217)

Or, similarly, in the *Theodicy* §44:

> [T]here are two great principles of our arguments. The one is the principle of *contradiction,* stating that of two contradictory propositions the one is true, the other false; the other principle is that of the *determinant reason*: it states that nothing ever comes to pass without there being a cause or at least a reason determining it, that is, something to give an *a priori* reason why it is existent rather than non-existent, and in this wise rather than in any other. (G VI 127/H 147)

We often take these pronouncements on Leibniz's part to suggest that PSR is, as it were, a fundamental axiom of his system, co-ordinate and co-equal to the Principle of Contradiction. And, on the standard story, based in large part on what immediately follows in both the *Monadology* and the *Theodicy*, the two principles govern two different sets of true propositions: the Principle of Contradiction applies to or undergirds truths of reason; the Principle of Sufficient Reason applies to and undergirds truths of fact. The Principle of Contradiction deals with necessary propositions; the Principle of Sufficient Reason deals with contingent propositions. While there are reasons to be dissatisfied with this first gloss (some of which will be discussed below), it supports what I take to be the unanalyzed view of the relation of the Principle of Contradiction and the Principle of Sufficient Reason: that they are, in some sense, independent of each other and *equally* fundamental. But I believe that there is still work to do. We ought to examine the relation between the two principles, and we ought to examine their foundations. Ultimately, we should ask what *justifies* or *guarantees* the truth of PSR, that is, we should ask what *grounds* the principle of the ground.[7] Such questions are very difficult, and Leibniz is not always particularly helpful; it is, indeed, quite rare for him to suggest just what grounds the PSR. On the one hand, this should not be surprising, for Leibniz holds that PSR is simply a fundamental feature of the world, something that guarantees the inherent intelligibility of the world. And he holds himself to be not so much the discoverer of the Principle of Sufficient Reason but as the philosopher who exposed a long-buried foundation of the world, of the human and divine intellect. On the

---

[7] Heidegger asks this: "Welches ist der Grund des Satzes vom Grund, von welcher Art ist dieser gewiß seltsame Grund?" (Heidegger 1957, 39).

other hand, he does offer a few suggestions of the ultimate grounds of the PSR and, in doing so, gives us *reasons* to believe in the Principle of Sufficient Reason. In this section, I want to examine these points.

While I shall concentrate on texts from his incredibly productive period in the 1680s, it is instructive to look at Leibniz's first attempt to provide an argument for the Principle of Sufficient Reason. It comes in a piece written in 1671–72, which has been given the title *Demonstratio Propositionum Primarum* by the Akademie editors and which was not published until 1966.[8] And while the proposition, *nihil est sine ratione*, is in its well-known form, Leibniz has not yet come to highlight this as one of the (two) fundamental principles of all our philosophizing.[9] It is "just" an important claim that stands in need of argumentative support. Leibniz writes the following:

> Proposition:
>
> *Nothing is without a reason*, that is, whatever is has a sufficient reason.
> Definition 1. *A sufficient reason* is that by virtue of which, if it is posited, a thing is.
> Definition 2. *A requisite* is that by virtue of which, if it is not posited, a thing is not.
>
> Demonstration:
>
> [1] Whatever is has all its requisites.
> For if one is not posited, the thing does not exist (by def. 2).
> [2] If all the requisites are posited, the thing exists.
> For if it does not exist, something will be lacking which keeps it from existing, that is, a requisite.
> [3] Therefore, all the requisites are a sufficient reason (by def. 1).
> [4] Therefore, whatever is has a sufficient reason. Q.E.D. (A VI ii 483)

This argument is weak, to say the least.[10] Implicit in Leibniz's argument is the view that will become explicit much later, that everything strives for existence and that, unless there is something to hinder the existence of some x, x will come into being. But a skeptic could claim that, even if all the necessary conditions (requisites) of x are present, it is still possible that x not exist. In other words, [2] really depends upon the Principle of Sufficient Reason and the thesis that all things (essences) strive for existence. As such, this argument is circular. And yet, Leibniz seems to have

[8]It is important for my later argument that some of Leibniz's texts were unknown to the philosophical world until recently.

[9]In a letter to Magnus Wedderkopf from May 1671, Leibniz does point to the importance of the thesis that nothing can exist without a sufficient reason for existing; without this thesis, he says, it is impossible to prove the existence of God and many other philosophical theses. (A II i 186) Still, we do not have *the Principle of Sufficient Reason*.

[10]This view is shared by Adams (1994, 68), who points to the apparent *petitio principii*, and Sleigh (CP 151, n.23), who refers to this "alleged proof."

thought highly of it—at least for a while. For it is repeated the following year in the *Confessio Philosophi* (A VI iii 118/CP 33) as well as his short piece *De Existentia* from December 1676:

> For existence, it is necessary that the aggregate of all requisites is present. A requisite is that without which a thing cannot exist. The aggregate of all requisites is the full cause of a thing. There is nothing without a reason; for there is nothing without an aggregate of all requisites." (A VI iii 587/DSR 111-13)

So, before the Principle of Sufficient Reason is elevated to the exalted status of one of the two great principles of all our reasonings, the young Leibniz *attempts* to provide a philosophical foundation for it. And even at the end of his career Leibniz appeals to the barebones of this view in his correspondence with Clarke.[11] But in his most thoughtful writings on the Principle of Sufficient Reason and his core metaphysical positions, this argument is not to be found. The reason, I think, is that Leibniz finds a much better argument.

Leibniz thought most intensely and most deeply about the core logical and metaphysical tenets of his system in the 1680s. As a result, it is in the texts of this period that we can find him engaged in an analysis and explication of the Principle of Sufficient Reason and a constellation of other fundamental principles of his philosophy. In particular, the short text that has come to be known to us as *Primary Truths* is a veritable treasure trove, containing *in nuce* most of his metaphysics. In my view, however, one of the most interesting features of Leibniz's *Primary Truths* is its argumentative structure. In fact, it is remarkable that this has actually drawn so little comment by interpreters of Leibniz's philosophy because it presents the Principles of Contradiction and Sufficient Reason *not* as co-equal principles of all our reasonings but rather as hierarchical, with the Principle of Contradiction or the Principle of Identity and his notion of truth as *grounding* in some sense the Principle of Sufficient Reason.[12] Given that the justification of the Principle of Sufficient Reason was an important problem among Leibniz's rationalist successors *and* given that this text was not published until 1903[13] and so unknown to these later philosophers, it is worth taking some time to try to understand exactly how Leibniz's argument proceeds here.

Leibniz gives a number of glosses on the Principle of Contradiction in his works from the 1680s:

1. For any proposition *p*, *p* is either true or false.
2. For any proposition *p*, *p* is not both true and false.
3. For any proposition *p*, if *p* implies a contradiction, then *p* is false.
4. For any proposition *p*, if *p* is false, then not-*p* is true.

---

[11] See Leibniz's Fifth Letter, §18. (G VII 393/L 698)

[12] This is discussed with subtlety by Carraud. See his (2002, 430f).

[13] C 518–23.

But, most interestingly, he also expresses this in the following way:

5. For any proposition *p*, if *p* is an identical proposition, then *p* is true.[14] (PC*)

All of these are on display in the short, opening paragraph of *Primary Truths*. And Leibniz concludes his account of the primary or first truths by saying that "they can all be included under the name 'identities'" (A VI iv 1644/AG 31). Another way to think of this, then, is that the Principle of Contradiction is a Principle of Identity. But the first stunning claim of this piece comes next: "all remaining truths are reduced to primary truths [that is, to statements of identity] with the help of definitions, that is, through the resolution of notions" (A VI iv 1644/AG 31). It will be precisely in this resolution of notions that so much work will be done. It becomes clear, for example, that the resolution of notions means a comparison of concepts, and it will be in this relation between concepts that Leibniz will see the very nature of truth: "the predicate or consequent is always in the subject or antecedent, and the nature of truth in general or the connection between the terms of a statement, consists in this very thing" (A VI iv 1644/AG 31). He goes on, "The connection and inclusion of the predicate in the subject is explicit in identities, but in all other propositions it is implicit and must be shown through the analysis of notions; a priori demonstration rests on this" (A VI iv 1644/AG 31). We can reformulate these theses as follows:

(T) All affirmative truths are either implicit or explicit identities.
(PIN) For any proposition *p*, *p* is true if and only if the predicate is in the subject (or the concept of the predicate is contained in the concept of the subject).[15]

Heidegger has a characteristically Delphic yet interesting way of putting this: "the essence of truth in general... lies in identity. Being true means being identical, *inesse* means *to be the same*."[16] In other words, in any true proposition, the concept of the predicate will be contained within the concept of the subject; but that also means that in the complete concept associated with some individual, there is some term that is *identical* to the term of the predicate.

[14] These formulations are taken from Sleigh (1983, 196). Leibniz's debt to Aristotle is great, and it is no surprise that Leibniz seeks to ground much of his philosophy on the Principle of Contradiction, just as Aristotle does in the *Metaphysics*. Compare the Stagirite's presentation of the Principle of Contradiction: "the same attribute cannot at the same time belong and not belong to the same subject and in the same respect," which is immediately glossed as "it is impossible for any one to believe the same thing to be and not to be" (*Metaphysics* 1005b18–19, 1005b23–24) .

[15] Leibniz also suggests that PIN has similar Aristotelian roots. Cf. *New Essays* 4.17: A VI vi 486/RB 486. Remnant and Bennett point to *Prior Analytics* 1.4 (25b32ff), where Aristotle argues that in syllogisms terms are "in" other terms. This seems to be what Leibniz has in mind with the conceptual containment of the predicate by the subject.

[16] "Wahrsein heißt Identischsein, inesse heißt *idem esse*." (Heidegger 1928, 49)

A notable consequence follows from T and PIN, something that Arnauld picked up on right away[17]: namely, it seems to render *all* truths *necessary* and to deny freedom to any action. In Arnauld's famous example, if Leibniz is right, then the concept of being celibate was contained in the individual concept of Arnauld; while it is true that Arnauld is celibate, it seems to make it impossible that he could have chosen to marry. More exactly, however, with this definition of truth, Leibniz renders all truth *analytic*—with one seeming exception: existence claims.[18] Now, I largely plan to ignore this can of worms and the details of Leibniz's account of contingency, that is, his theory of infinite analysis. But we need to recognize what Leibniz has in mind here with T and PIN. Take a simple proposition: "Alexander is a king." On the simple version of Leibniz's account, he is claiming that the concept of Alexander contains the concept of king. The truth of the proposition is precisely because of this internal relation of concepts; in other words, that's *why* it is a true proposition. But, in order to placate the kinds of concerns voiced by Arnauld, Leibniz must make the distinction between *necessary* and *contingent* propositions; and he does so in the following way:

> A true necessary proposition can be proved by reduction to identical propositions, or by reduction of its opposite to contradictory propositions; hence its opposite is called "impossible."
>
> A true contingent proposition cannot be reduced to identical propositions, but is proved by showing that if the analysis is continued further and further, it constantly approaches identical propositions, but never reaches them. Therefore it is God alone, who grasps the entire infinite in his mind, who knows all contingent truths with certainty. (*General Inquiries* (1686), §§133-34: A VI iv 776/ LLP 77)

The point is that, for any proposition, there is a relation between the concepts of the subject and predicate (conceived of as sets of properties) such that either through a finite or an infinite number of steps the sets of properties of the predicate are proven to be subsets of the properties of the subject.[19] As he says to Arnauld in the same year, "I do not intend any connection between the subject and the predicate other than that which holds in the most contingent of truths, that is, that we can always conceive something in the subject which serves to provide a reason why this predicate or event belongs to it. . ." (G II 46/AG 76). And this in turn leads us to the second of the "two great principles of all our reasoning": the Principle of Sufficient Reason. But the two points are connected, for one can argue that PSR properly conceived or taken to its logical consequences implies necessitarianism.

According to Leibniz, the Principle of Sufficient of Reason actually follows from T and PIN, that is, for his concept containment theory of truth. Let us return to his argument in *Primary Truths*, where this becomes explicit:

> [T]he received axiom that nothing is without a reason, or there is no effect without a cause [nihil esse sine ratione seu nullum effectum esse absque causa], directly follows from these

---

[17]See G II 64.

[18]See Russell (1937, 27).

[19]For more, see Sleigh (1982 and 1983) and Fitch (1979).

> considerations [PC*, T, and PIN]; otherwise there would be a truth which could not be proved a priori, that is, a truth which could not be resolved into identities, contrary to the nature of truth, which is always an explicit or implicit identity. (A VI iv 1645/AG 31)

In other words, given the fact that all truths are reducible to identities and are such that the concept of the predicate is contained in the subject, there must be a reason or an explanation for every truth. For if the Principle of Sufficient Reason were not true, then there would be an exception to the concept containment notion of truth: a truth for which there is no a priori proof that resolves into identities, a truth immune to analysis, be it finite or infinite. One might also explain this in the following way. If truth is solely a matter of the conceptual containment of the predicate in the subject, then God's omniscience entails knowing all such relations of containment. But to deny the Principle of Sufficient Reason means to claim that there is some occurrence or some truth that has no explanation; that is, there is a truth immune to explanation even by God. But that would mean that there is a truth that is not an instance of PIN.

When we look outside the *Primary Truths*, the connection between the Principle of Sufficient Reason and Leibniz's concept containment notion of truth becomes even more apparent. Consider the following passage from his July 1686 letter to Arnauld:

> [I]n every true affirmative proposition, whether necessary or contingent, universal or particular, the notion of the predicate is in some way included in that of the subject. *Praedicatum inest subjecto*, otherwise I do not know what truth is. [...] For *there must always be some foundation for the connection between the terms of a proposition, and this must be found in their concepts.* This is my great principle, with which I believe all philosophers should agree, and one of whose corollaries is the commonly held axiom that nothing happens without a reason, which can always be given, why the thing has happened as it did rather than in another way, even though this reason often inclines without necessitating. (G II 56/L 337)

Similarly, consider this, from the Appendix to the *Theodicy*:

> [T]here are two great principles, namely, *that of identicals or of contradiction*, which states that of two contradictory enunciations the one is true and the other false, and *that of the sufficient reason*, which states that there is no true enunciation whose reason could not be seen by one possessing all the knowledge necessary for its complete understanding. Both principles must hold not only in necessary but also in contingent truths; and it is even necessary that that which has no sufficient reason should not exist. For one may say in a sense that these two principles are contained in the definition of the true and false. (G VI 413-14/H 419)

It seems to me that Leibniz is as clear as he can be that the Principle of Sufficient Reason is a consequence of his concept containment notion of truth. Where Leibniz errs perhaps is in his belief that all philosophers should agree with this notion of truth—for clearly they do not and his case is not above reproach.

As mentioned above, the argumentative structure of the *Primary Truths* has received relatively little comment. We have gone through the opening paragraphs of this work and seen the following order of "discovery": (1) the Principle of Contradiction and the Principle of Identity; (2) the Predicate-in-Notion Principle; (3) the Principle of Sufficient Reason. But the argument of the *Primary Truths* actually points to the fact that PSR is dependent upon Leibniz's notion of truth and *not* on the Principle of Contradiction *simpliciter.* We should bear this fact in mind because Leibniz's successors attempt to derive the Principle of Sufficient Reason from the Principle of Contradiction, independently of a Leibnizian notion of truth.

We have now seen two attempts on Leibniz's part to ground the Principle of Sufficient Reason. It should be clear that Leibniz was committed to the Principle of Sufficient Reason *before* he adopted his concept containment theory of truth. I therefore agree with Adams, when he writes, "Leibniz would have believed in the Principle of Sufficient Reason even if he had never thought of the conceptual containment theory of truth. But the theory does provide an *explanation* of *how* there is a sufficient reason for every truth" (Adams 1994, 68).[20] In other words, Leibniz is guilty of a common philosophical sin: providing support for a thesis that he already believes in. But we should not think that Leibniz's concept containment theory of truth is some flimsy device to support a deep truth; it becomes a deep part of the Leibnizian metaphysics. Indeed, according to Sleigh (1983, 198), one of the most profound questions that we can ask about Leibniz's system is why he adopts the account of truth that he does. The simple answer—that doing so provides support for one of great principles of all our reasonings—must be rejected. There is more going on. A more substantial answer would begin by pointing to the fact that the divine intellect is a discursive intellect, an intellect that grasps the fundamental and intensional relations among concepts. It is only when truth is the conceptual containment of the predicate within the subject that (a) it is possible that there be non-existent possible worlds constituted by maximum sets of compossible essences—worlds in which it is true that Sherlock Holmes lived at 221b Baker Street or that Julius Caesar did not cross the Rubicon—and that (b) the existence of individuals in this world, that is, the actualization of this particular world, be the consequence of God's free decree. In other words, Leibniz's concept containment theory of truth follows from his commitments to divine omniscience, omnipotence and benevolence. And insofar as God is omniscient, omnipotent, and benevolent and his intellect is, like ours, discursive, the world is ordered and intelligible; that is, the Principle of Sufficient Reason follows from the nature of God as well. The similarity between the human and divine mind, then, is of great importance in the Leibnizian account of truth. As he writes in the *New Essays*, "it would be better to assign truth to the relationship amongst the objects of the ideas, by virtue of which one idea is or is not included within another. That does not depend on languages, and is something we have in common with God and the angels" (A VI vi 397/RB 397).

[20]The position that I am advocating here undermines the position of Couturat, for example, who essentially identified the Principle of Sufficient Reason with the concept containment theory of truth. See Couturat (1901, 208–21, esp. 214–15) See also Rauzy (2001, 51).

12.3 Leibniz's Rationalist Followers: Wolff and Baumgarten

Although Leibniz can be seen to argue for the Principle of Sufficient Reason and to attempt to provide some justification for this great principle of all our reasonings, for the most part this is not a project that concerned him. The Principle of Sufficient Reason was treated as simply an axiom of his system, immune from doubt and criticism. Indeed, Leibniz seems to have thought that his own particular achievement was to make PSR explicit and to draw all possible consequences from it in as a clear a manner as possible. His followers in the rationalist tradition in Germany, however, did see the lack of clear support for PSR as a potential weakness in metaphysics, and Wolff and Baumgarten, at least, sought to provide arguments for PSR that in one way or another depend upon the incontrovertible truth of the Principle of Contradiction. Since there has been little discussion of the details of the rationalist attempts to ground the Principle of Sufficient Reason in the literature, I would like to go through several of these arguments.

In one of the most important of his many, many writings, the *Ontologia*, Christian Wolff begins by explicating the "principles of first philosophy," with Chapter 1 devoted to the Principle of Contradiction and Chapter 2 the Principle of Sufficient Reason. His actual argument in §70 is relatively brief:

> Nothing is without a sufficient reason why it is rather than is not, that is, if something is supposed to exist, then something else is supposed by which it is understood why that thing is rather than is not. For either nothing is without a sufficient reason why it is rather than is not or something can be while lacking a sufficient reason why it is rather than is not (§53). Let us suppose that A is without a sufficient reason why it is rather than is not. Therefore, nothing is to be supposed by which it is understood why A is (§56). Indeed, it is admitted that A is because it is assumed that nothing is; but since this is absurd (§69), nothing is without a sufficient reason, or, if something is assumed to be, then something else is admitted by which it is understood why it is. (Wolff 1736, §70)

To better understand this argument, it will be helpful to look at the few claims from earlier sections. First, it is in §53 that Wolff gives us the Principle of Contradiction in the form of the Law of Excluded Middle: for any thing, it either is or is not (*Quodlibet vel est, vel non est*). And this in turn, Wolff tells us, coincides with another proposition of logic: of two contradictory propositions, one is necessarily true, the other false. Second, section 56 simply contains Wolff's definition of a sufficient reason: that by virtue of which it is understood why something is. And, finally, section 69 contains the crucial premise that, if nothing is assumed, it is not to be admitted that there is something. This argument can be reconstructed in the following way:

1. If it is assumed that nothing is, then it is not to be admitted that something is. (§69)
2. Either nothing is without a sufficient reason or something can exist without a sufficient reason. (By §59, the Principle of Contradiction, as the law of excluded middle: A or not-A.)
3. Assume A exists without a sufficient reason why it is rather than not.

4. Therefore, nothing is posited that explains why it is. (from §56, the def. sufficient reason)
5. But if it is admitted that A exists, then it is assumed that nothing exists (as its sufficient reason).
6. But this is absurd. For by (1) if it is assumed that nothing exists, then it is not to be admitted that something exists. And by *modus tollens*, if it is admitted that something exists, then it is not to be assumed that nothing exists.
7. Therefore, nothing is without a sufficient reason.

Here we have an argument that purports to derive the Principle of Sufficient Reason simply from the Principle of Contradiction and the definition of sufficient reason.

Yet, as with Leibniz's early argument for the Principle of Sufficient Reason, it is difficult to be moved by this argument. Indeed, this argument seems to rest on a logical mistake so elementary that it makes one blush: premise (1) and its converse allow Wolff to produce an ambiguity that makes his conclusion possible. After all, (1) need only be taken to mean "if there is nothing, then there is not something" and its converse "if there is something, then there is not nothing." But the argument depends upon our taking (5) as "if there is something, then there is nothing" and finding the contradiction with (1), when, in fact, we all know that what is meant is "if there is an A that has no sufficient reason, then there is nothing else outside of A that is its sufficient reason." Perhaps Wolff would say that in the moment of positing something, we necessarily posit something else. But that is simply to repeat the Principle of Sufficient Reason!

Our second derivation of the Principle of Sufficient Reason comes from the *Vernünftige Gedancken* of 1751 (or the *German Metaphysics* as it is commonly called). Wolff introduces PSR in §30, saying a number of things worthy of our attention.

> When something is present from which one can conceive why it is, that has a sufficient reason. Therefore, when nothing is present, there is nothing from which one can conceive why it is, namely, why it can actually be, and so it must come to be from nothing. Accordingly, that which cannot come to be out of nothing must have a sufficient reason why it is, as it must be possible in itself and have a cause which can bring it into actuality, if we are speaking of things that are not necessary. Now, since it is impossible that something can come to be out of nothing, everything that is must also have its sufficient reason why it is, that is, there must always be something by which one can understand why it can become actual. This proposition we wish to call the Principle of Sufficient Reason [*Satz des zureichenden Grundes*]. Mr. Leibniz proved the importance of this principle ... first in our times through beautiful examples [*Proben*] in his *Theodicy* as well as in the letters that he exchanged with the Englishman Clarke concerning several disputed points. He accepted it as a principle grounded in experience, to which there is no contrary case, and thus no demonstration of it, even though Clarke had demanded one.

It is important to note that Wolff appeals simply to the texts well-known to his audience, the *Theodicy* and the correspondence with Clarke; but neither here nor anywhere actually does he suggest a familiarity with the arguments that we examined in the previous section.[21] On the one hand, it might not be surprising that he believes that Leibniz held PSR to be a principle grounded in experience, for Leibniz does speak this way at times. But, on the other hand, even the *Theodicy* strongly suggests that the Principle of Sufficient Reason is not to be admitted as an empirical principle but rather as an innate and indubitable principle.

Be that as it may, the actual argument comes from the next section, §31, where Wolff writes the following:

> Suppose we have two things A and B that are identical (*einerley*). If something can be that has no sufficient reason why it is either in the thing (Sache) or outside of it, then a change can take place in A which does not occur in B, if one replaces B for A. In this sense B is not a thing identical with A. But since from the assumption that A is identical with B it follows that A is not identical with B if one does not allow the Principle of Sufficient Reason, it is on the contrary impossible that something can both be and not be at the same time. And so the same Principle has to have its unchallenged correctness. That is, it is true that everything has its sufficient reason why it is. (Wolff 1751)

This argument again appeals to the Principle of Contradiction—it is not the case that A and not-A—but in a different way. Here Wolff begins by assuming that we have two identical individuals and that the Principle of Sufficient Reason does not hold. If PSR does not hold, then something could take place in A that does not take place in B; in other words, if there were two indiscernible individuals and PSR did not hold, then A could develop in a way different from B. But, then, we would have $A = B$ and $A \neq B$, a violation of the Principle of Contradiction. Since this is impossible, Wolff tells us, we must conclude that it the Principle of Sufficient Reason must be true.

Leaving aside the confusion between identity and indiscernibility and Wolff's apparently tacit dependence upon the Principle of the Identity of Indiscernibles, there is a deeper reason behind the failure of this argument. In his *Dissertatio philosophica de usu et limitibus principii rationis determinatis vulgo sufficientis* (1743), Christian August Crusius claims that the Principle of Sufficient Reason can *never* be established from the Principle of Contradiction. According to Crusius, the Principle of Contradiction is an absolutely identical proposition; and so, insofar as it can be applied at all, it is necessary that the proposition be with respect to one and the same thing and one and the same *time*. But the Principle of Sufficient Reason as a statement of cause and effect presupposes different things (or qualities) at different times.[22] Since Wolff's argument from the *German Metaphysics* seems also to appeal to the possibility of change over time, we have very good reasons to reject it.

There is one last attempt to justify and establish the Principle of Sufficient Reason that I wish to consider here. It comes from one of Kant's favorite textbooks, the

---

[21] It is true that a condensed version of Leibniz's initial, superficial argument for PSR appears in his Fifth Letter to Clarke. But this shows mainly that the argument is so condensed that it did not make a sufficient impression on Wolff as he read the exchange.

[22] See Crusius 1743, esp. §XIV, in Vol. IV of Crusius (1969–87).

Metaphysica of Alexander Gottlieb Baumgarten, first published in 1739, but whose third edition of 1757 will be used here. Baumgarten presents the reader with a number of definitions and theses that will be at work in his argument for the Principle of Sufficient Reason: "nothing is A and not-A" or the Principle of Contradiction (§7); a possible is representable and is something that does not involve a contradiction (§8); "every possible is either A or not-A" or the Law of Excluded Middle (§10); "a reason [*ratio*] is that through it is understandable [*cognoscibile*] why something is" (§14) In §20, Baumgarten then writes the following:

> Every possible either has a reason [*ratio*] or it does not, §10. If it has a reason, then something is its reason, §8. If it does not have a reason, then nothing is its reason, §7. Therefore, every possible reason is either nothing or something, §10. If nothing had been the reason of some possible, it would have been understandable [*cognoscibile*] from nothing, why that thing is, §14, whence that very nothing is representable and something, §8; nothing [is] something, §14, 8. Hence, some possible is impossible, §7, 8, 9. Therefore, every possible reason is something, that is, every possible is a consequence [*rationatum*], that is, *nihil est sine ratione* or in positing something, something is supposed as its reason. *This proposition* is called *the principle of reason.* (Baumgarten 1757)

This argument is also frustrating, not least because the premises cited by Baumgarten do not exactly work in the way he suggests. Still, it should be clear that the argument depends largely on the Principle of Contradiction both in setting up the *reductio* and in the premise that a possible (a) implies no contradiction *and* (b) is representable. The failure of this argument, however, rests in precisely the second conjunct of this premise and Baumgarten's problematic reference to nothingness. For the crucial move is the argument is this: If *nothing* is the reason for some x, then that nothing is being represented as the reason for x. But whatever is representable is *something* and not nothing. Therefore, nothing is something, and we have generated the contradiction for the *reductio*. It is difficult to imagine being moved by this argument.

From this survey, it should be clear that Leibniz's rationalist successors, Wolff and Baumgarten, try to do what they believe Leibniz himself has never done: give an argument for the Principle of Sufficient Reason from the one truly indubitable foundation of our reasonings, the Principle of Contradiction. But as we have seen, these arguments ought to be considered embarrassing failures.[23] It should not escape our attention, however, that neither Wolff nor Baumgarten attempts to derive the Principle of Sufficient Reason from Leibniz's concept containment theory of truth. On the one hand, this should not surprise us: the texts in which Leibniz enunciates this doctrine most clearly were unknown to Wolff and Baumgarten. On the other hand, the fact that neither seems to believe that truth consists in the inclusion of

---

[23]The young Kant gives an argument for a special form of the Principle of Sufficient Reason. In the *Nova dilucidatio* (1755), he argues for his Proposition VIII: "Nothing which exists contingently can be without a ground which determines its existence antecedently." (AA I 396) This argument, however, is *not* an embarrassing failure and gets at a number of deep issues. I plan to treat this on another occasion.

the concept of the predicate within the concept of the subject shows that Leibniz's certainty in this as a definition of truth might be misplaced.

12.4 Hume

As Leibniz usually presents it, the Principle of Sufficient Reason is simply a principle or axiom of all our reasonings about contingent things, which is intuitively certain. In other words, it derives its force primarily from the fact that we cannot imagine an effect lacking a cause. Now, as all students of philosophy know, David Hume argued against the idea of a necessary connection between events and against the general principle that every effect has a cause. Indeed, Hume's position on this matter in the *Treatise of Human Nature* is two-fold: first, that the general causal principle cannot be considered intuitively certain and, second, that there is no argument that can demonstrate this.[24] Now, insofar as the general causal principle is entailed by the Principle of Sufficient Reason, his argument can have obvious implications for PSR itself. It is interesting, however, that the proponents of the Universal Causal Principle who draw Hume's ire in the *Treatise of Human Nature* are typically considered "empiricists": Hobbes, Clarke, and Locke. And it is against *their* arguments that Hume directs his attack. Let us look briefly at Hume's arguments and by extension to the original arguments from Hobbes and Locke.[25] To what extent do these arguments differ from those that we have seen thus far? Could any of the rationalist arguments withstand his critique? Again, according to Hume, "every demonstration, which has been produc'd for the necessity of a cause, is fallacious and sophistical" (*THN* 1.3.3.4).

In the crucial section the *Treatise*, "Why a cause is always necessary" (*THN* 1.3.3), Hume analyzes three arguments that purport to demonstrate the Universal Causal Principle. His first target is an argument attributable to Hobbes.[26] Imagine that something were to arise without a cause. Since all points of time and space are alike, if there were no cause, then the thing could not come into being. That is, there

[24]As this paper concerns the arguments for, or purported demonstrations of, the Principle of Sufficient Reason, I shall pass over the topic of its intuitive certainty. But I do wish to make one remark. It was claimed at the beginning of this paper that the Principle of Sufficient Reason is a fundamental tenet of "rationalism". While this is true, it is more accurate to say that "rationalism" holds the Principle of Sufficient Reason to be innate and, therefore, intuitively certain. It is for this reason that Descartes, Spinoza, and Malebranche provide no argument for it, nor highlight it in the way that Leibniz and his followers do. What I wish to emphasize here, however, is that Leibniz, Wolff, Baumgarten, and the pre-critical Kant consider the Principle of Sufficient Reason to be logically dependent upon other, more fundamental, notions. Compare this idea with that of Heidegger, who claims that the Principle of Sufficient Reason is the most basic of all principles; the Principle of the Ground is the Ground of all Principles, we are told (Heidegger 1957, 30–31).

[25]Hume attributes the second argument to Samuel Clarke, but it is not clear that Clarke actually favored such an argument. See Hume (1739–40/2007), vol. 2, 736.

[26]From *On Liberty and Necessity* in *English Works* 4:276.

is nothing about the nature of space and time that can explain for the coming-to-be of some thing at some particular time or at a particular place. But, Hume claims, we can suppose that the time and place are fixed without a cause just as easily as we can suppose that the thing's existence is fixed without a cause; indeed, according to Hume, there are two separable questions: first, *whether* something comes into existence and, second, *when* and *where* that thing comes into existence. But if there is no intuitive absurdity in claiming that something comes into being without a cause, then there should be no intuitive absurdity with saying that that it comes into existence at some time or place; and the seeming absurdity of the one claim cannot prove the absurdity of the other, "since they are both upon the same footing, and must stand or fall by the same reasoning" (*THN* 1.3.3.4). We ought not to have any trouble concluding with Hume that this argument is a failure. But we should note that it is a *new* kind of failure, unlike any of the arguments that we saw from Leibniz and his followers.[27]

The second argument is attributed by Hume to "Dr. Clarke and others" and is simply this: if something were to lack a cause, it would have to be a cause of itself, that is, it would have to exist before it existed; since this is impossible, nothing (contingent) can exist without a cause. But, Hume argues, in denying that there must be causes for all things, he is denying even that a thing itself is to be considered a cause of itself. For if one claims that, lacking a cause outside of itself, a thing must be a *cause* of itself, then one is simply assuming the very causal principle that Hume questions.

The third argument complements the second and fails for the same reason. In Hume's words, it runs as follows: "Whatever is produc'd without any cause, is produc'd by *nothing*; or in other words, has nothing for its cause. But nothing can never be a cause, no more than it can be something..." (*THN* 1.3.3.6) Now Hume points to Locke for this argument, when, in fact, Locke claims that it is intuitively certain that bare *nothing can no more produce any real Being, than it can be equal to two right angles* (Locke, *Essay* IV.10.3). Thus, this "demonstration" is actually Hume's invention based on Locke's text. Nevertheless, it is obvious that it bears a strong similarity to the arguments that we saw produced by the rationalists, Wolff and Baumgarten. The failure of this argument is again, according to Hume, that it begs the question. We suppose that something has no cause, and we are then told that *nothing* must be the cause of it, which is impossible, since *nothing* can produce no effects. As Hume writes,

> If every thing must have a cause, it follows, that upon the exclusion of other causes we must accept of the object itself or of nothing as causes. But 'tis the very point in question, whether every thing must have a cause or not; and therefore, according to all just reasoning, it ought never to be taken for granted. (*THN* 1.3.3.7)

---

[27]One might argue that it bears a similarity to the argument that Leibniz advances in his correspondence with Clarke for the Principle of the Identity of Indiscernibles, which is based on the Principle of Sufficient Reason. Cf. G VII 363–64/AG 325.

The dissimilarity between this argument and the arguments we saw in the previous section is that it is not presented as some kind of consequence of the Principle of Contradiction.

We have now considered a number of arguments for the Principle of Sufficient Reason, all of which have been revealed to be manifest failures, with one exception: Leibniz's argument in his metaphysical and logical writings of the 1680s as exemplified in the *Primary Truths*, in which the Principle of Sufficient Reason is shown to follow from the Principle of Contradiction and the Leibnizian theory of truth. These texts were, of course, unknown to Hume. But what might Hume or a Humean say in response? First, Hume has a different definition of truth: "Truth is of two kinds, consisting either in the discovery of the proportions of ideas, consider'd as such, or in the conformity of our ideas of objects to their real existence" (*THN* 2.3.10.2). Truth is *not*, as it is for Leibniz, simply the conceptual entailment of the predicate by the subject. Moreover, according to Hume, knowledge only arises from relations of ideas, but the ways in which ideas can be related compared are limited (*THN* 1.3.3). And it is fair to say that Hume would not recognize the Leibnizian notion of conceptual containment that lies at the foundation of the Leibnizian theory of truth. Indeed, this will have parallels in the Amphiboly chapter of Kant's *Critique of Pure Reason*.[28] There Kant criticizes Leibniz for "intellectualizing the appearances" and for failing to distinguish between noumena and phenomena. And the point here is that the way in which Leibniz claims that we can compare concepts and judge one to be contained within another is, for both Hume and Kant, illegitimate.[29]

There is, however, a sympathetic Humean reading of Leibniz's theory of truth that will help us go slightly deeper into the issue. On Hume's view, knowledge can arise from the *resemblance* and *contrariety* of ideas; that is, we can know immediately that two objects resemble each other (*THN* 1.3.1.2). Now, as was discussed above, Leibniz holds that the proposition "Alexander is a king" is true because the concept *Alexander* contains the concept *king*. But this in turn means that the complete individual concept of Alexander contains either *king* itself or certain marks attendant to being a king. A Humean straining with near superhuman sympathy might say that the concept containment theory of truth amounts to perceiving the resemblance between the concept of the predicate within the concept of the subject or between the marks attendant to being a king that are the concept of the predicate and those "same" marks within the concept of the subject. This is, indeed, near superhuman interpretive sympathy, for a genuine Humean would certainly dig in his heels at the idea of our perceiving *concepts*, especially given Hume's rejection of abstract ideas (*THN* 1.1.7). In the end, then, Hume or even his kindest follower could

[28] AA III 219–33, that is, A 260–92/B 316–49; see especially AA III 228, or A 284/B 340.

[29] Although Leibniz writes that ideas can be contained in other ideas in the *New Essays*, we ought to hold this for a slip of the pen; for strictly speaking it is *concepts* and not ideas that are contained within others.

not be persuaded by Leibniz's explanation of the Principle of Sufficient Reason, given its metaphysical and logical foundations.

Does all of this mean that Hume *rejects* the Universal Causal Principle, or even the Principle of Sufficient Reason? No. In a letter to John Stewart, he is quite explicit: "allow me to tell you, that I never asserted so absurd a Proposition as *that any thing might arise without a Cause*: I only maintain'd, that our Certainty of the Falshood of that Proposition proceeded neither from Intuition nor Demonstration; but from another Source."[30] For our purposes, the crucial point is that Hume allows for the truth of the Principle of Sufficient Reason while rejecting all demonstrations of it. This is important, for one can occasionally encounter a fellow philosopher who will speak disparagingly of the Principle of Sufficient Reason and express comfort with "brute facts." Compare this assertion with Hume's claim that "what the vulgar call chance is nothing but a secret and conceal'd cause" (*THN* 1.3.12.1). Now, we have been taught that there are phenomena at the quantum level that are immune to causal explanation; at this level, then, there might be "brute facts." And if this is the case, then certainly the Principle of Sufficient Reason cannot be considered universal and necessary. But, in all other cases, we should probably not give up the Principle of Sufficient Reason for the metaphysical machismo that asserts the existence of "brute facts" throughout the world. Following Hume *and* Leibniz, we might reasonably assert that the facts that present themselves to us as "brute" follow from secret and concealed causes.

12.5 Conclusion

In tracing the development of arguments for the Principle of Sufficient Reason, we have seen that Leibniz is unique among the rationalists. For Leibniz, the Principle of Sufficient Reason is part and parcel of his conception of truth, and this in turn is a consequence of his conception of God, of God's intellect, and the nature of possibility. Further, the Leibnizian theory of truth and the Principle of Sufficient Reason are founded on Leibniz's commitment to a similarity between the human and divine intellects. This is of great importance, as it guarantees the intelligibility and rationality of the world. Of course, the complete reasons for contingent features of the world will surpass the human intellect, but we can still be certain that there *are* reasons. Naturally, if one denies this conception of the nature of God, one can easily deny this account of the nature of truth; and if one denies this account of the nature of truth, one can reject the Principle of Sufficient Reason. On the other hand, if one denies the Principle of Sufficient Reason, one *must* reject the conception of truth that Leibniz finds most obvious and natural as well as much of a traditional theistic conception of God.

I concluded the previous section by showing that not even Hume believes that the Universal Causal Principle or the Principle of Sufficient Reason is *false*—only that

---

[30] Feb. 1754, Letters 1:186.

there can be no demonstrations of its truth. While we saw that certain "empiricists" offered arguments for the Universal Causal Principle, if we were to look for a signal feature of Leibnizian "rationalism", it might simply be in the commitment to the demonstrability of this fundamental feature of the human mind and world. Hobbes and Locke, for example, are actually acting in a rationalist spirit in giving those arguments that Hume critiques. Thus, in "rationalism" there is a continuity with the tradition of western metaphysics insofar as the Principle of Sufficient Reason in any of its many guises can be traced back to Parmenides, Aristotle, and Archimedes, and buttresses so many of the scholastic arguments for the existence of God. And in *Leibnizian* rationalism there is an important *innovation*: the attempt to deduce the Principle of Sufficient Reason from the Principle of Contradiction. It is perhaps, however, only Leibniz himself who, in attempting to ground the Principle of Sufficient Reason in his account of truth and the consequences of his theism, is at the same time traditional and innovative. But does this violate the Principle of Contradiction?

References

Adams, Robert M. 1994. *Leibniz: Determinist, theist, idealist*. Oxford: Oxford University Press.
Aristotle. 1984. *The complete works of Aristotle*. Princeton, NJ: Princeton University Press.
Baumgarten, Alexander. G. 1757. *Metaphysica*. Halle: Hemmerde.
Carraud, Vincent. 2002. *Causa sive ratio: La raison de la cause, de Suarez à Leibniz*. Paris: Presses Universitaires de France.
Couturat, Louis. 1901. *La logique de Leibniz d'après des documents inédits*. Paris: Félix Alcan.
Crusius, Christian. A. 1969–87. *Die philosophischen Hauptwerke*. Hildesheim: Olms Verlag.
Descartes, René. 1964–74. *Oeuvres*. Paris: Vrin.
Descartes, René. 1985. *The philosophical writings of Descartes*. Cambridge: Cambridge University Press.
Fitch, Gregory. 1979. Analyticity and necessity in Leibniz. *Journal of the History of Philosophy* 17: 29–42.
Heidegger, Martin. 1928. *Metaphysische Anfangsgründe der Logik im Ausgang von Leibniz*. Frankfurt: Vittorio Klostermann.
Heidegger, Martin. 1957. *Der Satz vom Grund*. Stuttgart: Klett-Cotta.
Hume, David. 1739–40/2007. *A treatise of human nature*. Oxford: Clarendon Press.
Hume, David. (1932). *The letters of David Hume*. Oxford: Clarendon.
Kant, Immanuel. 1902–1983. *Kants Gesammelte Schriften, herausgegeben von der Deutschen Akademie der Wissenschaften*. Berlin: De Gruyter.
Leibniz, Gottfried W. 1875–90. *Die philosophischen Schriften von Gottfried Wilhelm Leibniz*. Berlin: Weidman.
Leibniz, Gottfried W. 1903. *Opuscules et fragments inédits de Leibniz*. Paris: Félix Alcan.
Leibniz, Gottfried W. 1923-. *Sämtliche Schriften und Briefe*. Berlin: Akademie Verlag.
Leibniz, Gottfried W. 1966. *Logical papers*. Oxford: Clarendon Press.
Leibniz, Gottfried W. 1969. *Philosophical papers and letters*. Dordrecht: Reidel.
Leibniz, Gottfried W. 1981. *New essays on human understanding*. Cambridge: Cambridge University Press.
Leibniz, Gottfried W. 1985. *Theodicy*. La Salle, IL: Open Court.
Leibniz, Gottfried W. 1989. *Philosophical essays*. Indianapolis: Hackett.
Leibniz, Gottfried W. 1992. *De Summa Rerum*: Metaphysical Papers, 1675–76. New Haven, CT: Yale University Press.

Leibniz, Gottfried W. 2005. *Confessio Philosophi: Papers concerning the problem of evil, 1671–1678*. New Haven, CT: Yale University Press.
Malebranche, N. 1958–68. *Oeuvres complètes*. Paris: Vrin.
Pruss, A. R. 2006. *The Principle of sufficient reason: A reassessment*. Cambridge: Cambridge University Press.
Rauzy, J.-B. 2001. *La doctrine leibnizienne de la vérité: Aspects logiques et ontologiques*. Paris: J. Vrin.
Russell, B. 1937. *A critical exposition of the philosophy of Leibniz*. London: Allen & Unwin.
Sleigh, R. C., Jr. 1982. Truth and sufficient reason. In *Leibniz: Critical and interpretive essays*, ed. M. Hooker, 209–42. Minneapolis, MN: University of Minnesota Press.
Sleigh, R. C., Jr. 1983. Leibniz on the two great principles of all our reasoning. *Midwest Studies in Philosophy* 8: 193–216.
Spinoza, B. 1985. *The collected works of Spinoza*. Princeton, NJ: Princeton University Press.
Wolff, C. 1736. *Philosophia Prima, sive Ontologia*. Frankfurt: Renger.
Wolf, C. 1751. Vernünftige Gedancken von Gott, der Welt, und der Seele des Menschen. Halle: Hemmerde.

Name Index

C. Fraenkel et al. (eds.), *The Rationalists: Between Tradition and Innovation*,
The New Synthese Historical Library 65, DOI 10.1007/978-90-481-9385-1,

Subject Index

Papier | Fördert gute Waldnutzung
FSC® C083411

Kolibri 360 GmbH
Ferdinand-Jühlke-Straße 7
99095 Erfurt, Deutschland
produktsicherheit@kolibri360.de